EXPERIMENTS IN TEXTILE AND FIBRE CHEMISTRY

Experiments in Textile and Fibre Chemistry

Christopher Earland,
M.Sc., Ph.D., A.R.G.S., F.R.I.C., F.T.I.
and
David J. Raven,
M.Sc., Ph.D., A.R.I.C.

*School of Textile Science and Technology,
Bradford University*

LONDON

BUTTERWORTHS

THE BUTTERWORTH GROUP

ENGLAND
Butterworth & Co (Publishers) Ltd
London: 88 Kingsway WC2B 6AB

AUSTRALIA
Butterworth & Co (Australia) Ltd
Sydney: 20 Loftus Street
Melbourne: 343 Little Collins Street
Brisbane: 240 Queen Street

CANADA
Butterworth & Co (Canada) Ltd
Toronto: 14 Curity Avenue, 374

NEW ZEALAND
Butterworth & Co (New Zealand) Ltd
Wellington: 49/51 Ballance Street
Auckland: 35 High Street

SOUTH AFRICA
Butterworth & Co (South Africa) (Pty) Ltd
Durban: 33/35 Beach Grove

First published 1971

Suggested U.D.C. No. 542/543:677

ISBN 0 408 70089 0 Standard

0 408 70102 1 Limp

Filmset and Printed in England by
Page Bros. (Norwich) Ltd.,
Norwich and London.

Preface

This book describes fifty-one selected experiments in the chemistry of fibrous polymers and ancillary materials which have been designed primarily for undergraduate students in Technical Colleges, Polytechnics and Universities. Since the division between textiles and polymers is somewhat arbitrary, the book is aimed at students who are studying either of these technologies. The experiments vary in difficulty, and whereas the simpler ones could be performed by senior pupils in schools, some are of a research nature and are suitable for post-graduate courses in polymer or textile science. Since natural fibres are inherently variable and also require purification before use which may cause appreciable chemical modification, commercially they are subjected to more chemical tests than the synthetics. This is reflected in the relative numbers of experiments devoted to the two classes of fibres.

Many of the experiments are original and all have been performed many times in the authors' laboratories to ensure that they will produce worth-while results. As a matter of policy, experimental procedures are given in detail to avoid the necessity for students to consult reference books which may not be readily available during practical periods. It is not intended, however, that these experiments should be regarded merely as exercises in handling apparatus and materials. It is important that they be read as a whole and the principles on which they are based are clearly understood before the commencement of practical work. Where possible, the references cited, which range from general articles to original papers, should be consulted and if appropriate the information may be incorporated in the laboratory report.

The authors wish to thank their colleagues Mr. J. T. Ball and Mr. D. E. Montgomery for helpful suggestions and Dr. I. Smith and Dr. J. G. Feinberg and the Shandon Scientific Company Ltd., for permission to use material contained in their book *Paper and Thin Layer Chromatography and Electrophoresis*.

Finally, we would like to acknowledge the help of Mrs. C. Earland in the preparation of the manuscript and Mr. W. McDonald for assistance in checking many of the experiments.

CHRISTOPHER EARLAND
DAVID J. RAVEN

Bradford

Contents

General notes
Determination of dry weight of fibrous polymers

PART 3. *Experiments on Carbohydrates*

PART 4. *Experiments on Synthetic Polymers*

Appendix

General notes

The apparatus listed excludes common items which are normally found in a chemical laboratory. The list of materials, however, includes all chemicals required for the experiment except concentrated hydrochloric, nitric and sulphuric acids, 0·880 ammonia solution and solid sodium hydroxide. Where possible, the reagents used should be of 'Analar' grade. If a chemical is not manufactured in this grade or it is costly, 'Laboratory Reagents' (BDH) or 'Laboratory Chemicals' (M & B) may be used. Where ethyl alcohol is specified for an experiment, anhydrous (74° over proof) Industrial Methylated Spirits (IMS) is satisfactory.

If a chemical is excessively toxic or an operation requires special safety precautions, this is indicated in the text. Particular attention, however, should be paid to the following.

Sealed tubes containing acid hydrolysates must in no circumstances be handled when hot. The sealed tube should be placed upright in a small glass beaker contained in a stainless steel beaker of adequate height and the whole placed in the oven. It is permissible to open the oven during the heating operation, but the tubes must not be disturbed nor removed until the oven is cold.

Glass vacuum desiccators should be of heavy construction and obtained only from a reputable manufacturer. To avoid 'sucking back' due to variation of the water pressure, water pumps should be fitted with a non-return device or a filter-flask inserted in the line between the pump and the desiccator.

The accuracy to which quantities should be measured is indicated by the number of decimal places given, e.g. 2 g implies a mass between 1·5 and 2·5 g. 2·0 g a mass between 1·95 and 2·05 g and 2·00 g a mass between 1·995 and 2·005 g. Similarly, 25·0 ml should be within the range 24·95 and 25·05 ml. For volumetric analysis, grade

B apparatus is sufficiently accurate. Small volumes, even for preparative work, are conveniently measured by graduated pipettes rather than by measuring cylinders.

Determination of dry weight fibrous polymers

Although all substances are to some extent hygroscopic, the water content of many inorganic and organic chemical reagents is so small that it may be disregarded. In the case of fibrous polymers, however, their water content under normal atmospheric conditions is often appreciable and to obtain their correct weight, they must either be weighed after drying or a correction factor must be employed.

NOMENCLATURE

If a substance is said to contain x per cent of water (or moisture) this is normally taken to mean that it contains x parts by weight of water and (100-x) parts of dry substance. This interpretation is accepted also in the textile industry where the quantity x is known as the *moisture content* or *direct loss*.

In the textile industry, however, a second and more usual method of expressing water content is employed. This is the *regain*, which is the quantity of water present expressed as a percentage of the *dry* weight of the substance.

If a weight (w) of dry material has a regain of y per cent, then the following relationships hold:

$$\text{Weight of water present} \quad = \frac{wy}{100} \tag{i}$$

$$\text{Weight of moist material } (w_1) = w + \frac{wy}{100} \tag{ii}$$

$$= w(1 + 0{\cdot}01y) \tag{ii}$$

$$\text{Percentage moisture content} = 100 \left[\frac{wy}{100} \div w(1 + 0{\cdot}01y) \right]$$
$$\text{or direct loss } (x)$$
$$= \frac{y}{(1 + 0{\cdot}01y)} \tag{iii}$$

From equation (iii) it follows that (*a*) the moisture content or direct loss must always be less than the regain and (*b*) as the water uptake decreases, the regain and moisture content approach one another in value.

BALANCES

Three types of balance are required to carry out the experiments listed. (*a*) Scales with a capacity of up to 100 or 200 g and reading to 0·1 g, (*b*) a rough balance with dial and chain attachment reading to 0·01 g and (*c*) an analytical balance reading to 0·0002 g. Clean and dry fibrous materials may be weighed directly on the pans of (*a*) and (*b*), but when using an analytical balance a suitable container must always be employed.

For analyses on very small quantities of materials, for example a few fibres, a semi-micro balance reading to 0·02 mg is required, but such a balance is not essential for the procedures given.

CHOICE OF DESICCANT

Three agents adequately cover all the needs for drying the materials mentioned in this book.

1. Self-indicating silica gel is a general purpose drying agent. It is particularly useful for desiccators when they are used for allowing substances to cool in prior to weighing. It may be used also for removing the bulk of the moisture from fibrous materials but it will not dry them completely if they have a high regain. When silica gel becomes pink, it should be heated in an oven at 110–120°C until the original blue colour is restored.

2. Phosphorus pentoxide (phosphoric oxide) is an extremely effective drying agent and for practical purposes will dry completely all fibrous materials. Its disadvantages are that care is necessary in its handling and it is rendered liquid and useless by a small amount of absorbed water. It cannot be regenerated by drying and any

unwanted material should be destroyed, in small quantities at a time, by the rapid addition of a large volume of cold water. Whereas silica gel may be placed directly in the bottom of a desiccator, for economy and ease of cleaning, phosphorus pentoxide should be placed first in a dish. It is often advantageous to dry a material first over silica gel and to complete the process over phosphorus pentoxide.

3. Solid sodium hydroxide is an excellent agent for drying hydrolysates containing hydrochloric acid as it removes both acid and water. The hydrolysate is placed in a dish which is contained in a desiccator holding the desiccant in the form of flakes or pellets.

In all these procedures, the rate of drying may be accelerated by evacuating the desiccator with a water pump (for precautions *see* page ix). The use of calcium chloride as a desiccant is not recommended.

DETERMINATION OF DRY WEIGHT

Although slovenly work must always be avoided, there is no point in determining the weight of a material with a degree of accuracy far in excess of the requirements of the experiment. Depending upon the accuracy required, three methods may be employed for ascertaining the dry weight of a fibrous material.

(*a*) If a material has been exposed to the atmosphere for a sufficient time, its moisture content will be in equilibrium with atmospheric conditions. If the value of the 'regain' (y) of the material is known for these conditions, then it follows from equation (ii) on page xi that if $(1 + 0{\cdot}01y)$ is denoted by f, it will be necessary to weigh wf grammes of material if w grammes of dry material are required.

Values for 'f' applicable to the Authors' laboratories are given in the table.

Although this method is valuable for semi-quantitative work, it has obvious limitations.

(*b*) A sample of the material (1–2 g) is weighed (w_1) and dried for 1 h at 110°C. The dry weight is w. Since $w_1/w = f$, it is necessary only to weigh the remaining samples without drying and divide by the factor f, as in method (*a*).

This method is a more accurate application of method (*a*) and provided f is determined daily on a representative sample, is suitable for all but the most accurate work.

(*c*) For the most accurate analytical work there is no substitute for the direct weighing of the dry material. The material is spread out in an oven for 1 h at 110°C and weighed after transferring to a

Material	$f = (1 + 0{\cdot}01y)$
Terylene	1·00
nylon 66	1·04
cellulose triacetate	1·05
cellulose diacetate	1·06
cotton	1·09
viscose	1·11
silk	1·11
wool	1·16

weighing bottle which is allowed to cool in a desiccator. The weight of the bottle has been determined previously after heating and cooling under the same conditions. Should the material be sensitive to heat, it is dried in an evacuated desiccator over phosphorus pentoxide for 2 days. These drying times are applicable to atmospherically dry material. Wet substances should be dried for 2 h in the oven or pre-dried over silica gel before drying over phosphorus pentoxide.

Part 1.   Analytical Methods

1 Determination of 'available' chlorine in sodium hypochlorite solution

APPARATUS

Only normal volumetric apparatus is required.

MATERIALS

Potassium iodate, potassium iodide, sodium thiosulphate, sodium hypochlorite solution (commercial or laboratory reagent quality (*see* Note 1).

THEORY

If aq. chlorine be required in industry, it is generally obtained in the form of sodium hypochlorite solution produced by the reaction of the element with alkali

$$2NaOH + Cl_2 = NaOCl + NaCl + H_2O$$

The composition of this solution is dependent on its pH (for details *see* Alexander *et al* [1]) but may be summarised as follows:
In strongly acid solution the chlorine is present in the elemental form

$$NaOCl + 2H^+ + 2Cl^- \rightarrow NaCl + H_2O + Cl_2$$

In weakly acid solution, hypochlorous acid is present

$$NaOCl + H^+ + Cl^- \rightarrow NaCl + HOCl$$

3

whilst in alkaline solution, hypochlorite ions predominate

$$NaOCl \rightarrow Na^+ + OCl^-$$

It should be noted that all these products oxidise two molecules of potassium iodide to a molecule of iodine, although the number of chlorine atoms involved in the left-hand side of the equations may differ.

$$2KI + Cl_2 \rightarrow 2KCl + I_2$$
$$2KI + HOCl \rightarrow KCl + KOH + I_2$$
$$2KI + OCl^- \rightarrow K_2O + Cl^- + I_2$$

The 'available chlorine' is the oxidising power expressed as chlorine and is given by the volumetric relationship.

$$1000 \text{ ml N } Na_2S_2O_3 \text{ soln.} \equiv 35.5 \text{ g available chlorine}$$

It should be appreciated that a substance may contain more than its own weight of 'available' chlorine.

EXPERIMENTAL PROCEDURE

(a) *Preparation of standard* 0.1 N-*potassium iodate*—Potassium iodate reacts with potassium iodide in acid solution to produce iodine

$$KIO_3 + 5KI + 3H_2SO_4 = 3K_2SO_4 + 3H_2O + 3I_2$$

$$\therefore \text{Equivalent wt. } KIO_3 = \frac{\text{Molecular weight}}{6} = 35.67$$

Weigh accurately between 0.85 and 0.95 g of A.R. potassium iodate (w g) on a small watch glass and transfer, using a funnel and distilled water, to a 250 ml graduated flask. Dissolve in distilled water in the flask and make up to the mark. Shake well to ensure uniform mixing. Calculate the exact normality of the iodate solution.

$$\text{Normality} = \frac{4w}{35.67}$$

(b) *Preparation and standardisation of* 0.1 N-*sodium thiosulphate* (*see* Note 2)—Dissolve sodium thiosulphate (25 g) and borax (3.8 g), which acts as a preservative, in distilled water in a 1 litre graduated flask. Make up to the mark and shake well.

By means of a pipette, transfer 25 ml portions of the standard potassium iodate solution to a 250 ml conical flask. Add 10 per cent potassium iodide solution (approx. 10 ml) and 2N sulphuric acid (approx. 5 ml) and titrate the liberated iodine with the sodium thiosulphate solution from a burette. Take the mean of three concordant titrations (T ml) and calculate the normality of the sodium thiosulphate solution.

$$\text{Normality } (x) = \left(\frac{25 \cdot 0}{T}\right) \times \text{Normality KIO}_3 \text{ solution}$$

(*c*) *Determination of available chlorine in sodium hypochlorite solution*—Pipette 10 ml of the conc. sodium hypochlorite solution into a 250 ml graduated flask and make up to the mark with distilled water. Shake to ensure uniform mixing. By means of a pipette transfer 25 ml portions to a 250 ml conical flask. Add potassium iodide solution and acid and titrate the iodine with sodium thiosulphate solution using the procedure described under (*b*). If the mean of three concordant titrations is y ml.

since, $1 \cdot 0$ ml N $Na_2S_2O_3$ solution $\equiv 0 \cdot 0355$ g   available chlorine

then, y ml xN $Na_2S_2O_3$ solution $\equiv 0 \cdot 0355xy$ g available chlorine

$\therefore$ 25 ml of diluted hypochlorite contain $0 \cdot 0355xy$ g of available chlorine and the conc. hypochlorite contains $3 \cdot 55xy$ g of available chlorine per 100 ml.

NOTES

1. Sodium hypochlorite solution as manufactured generally contains a maximum of 15 per cent available chlorine. It should be kept in a cool place in the dark with the bottle closed by a rubber cap. Weekly standardisation is essential as it may deteriorate rapidly.

2. A.R. Sodium thiosulphate contains at least 99 per cent $Na_2S_2O_3 \cdot 5H_2O$ (mol. wt. $248 \cdot 2$) and this grade may be used as a primary standard ($25 \cdot 0$ g per litre) for all but the most accurate work.

REFERENCE

1 ALEXANDER, P., HUDSON, R. F. and EARLAND, C., *Wool—Its Chemistry and Physics,* Chapman & Hall, London. 272 (1963).

2 Determination of the hardness of water

APPARATUS

No special apparatus is required.

MATERIALS

Standard soap solution (Wanklyn's), 0·02N-ethylene diamine tetra-acetic acid (EDTA), ammonia buffer solution, total hardness and calcium hardness indicator tablets (all these reagents may be obtained from B.D.H.),* 4N-sodium hydroxide (16 g solid made up to 100 ml with water), samples of water (*see* Note 1).

THEORY

Hard water destroys soap due to the presence of soluble calcium and magnesium salts, for example

$$2C_{17}H_{35}COONa + Ca^{2+} \rightarrow (C_{17}H_{35}COO)_2Ca \downarrow + 2Na^+$$

Temporary hardness is due to the presence of calcium or magnesium bicarbonates and can be removed by boiling.

$$Ca(HCO_3)_2 \rightarrow CaCO_3 \downarrow + H_2O + CO_2$$

Permanent hardness cannot be removed by boiling, since the soluble salts of magnesium or calcium present are stable to this treatment, for example sulphates and chlorides.

A discussion of the reagents employed in these determinations is

* Poole, Dorset, England.

given by Vogel [1]. It should be noted that hardness is generally expressed as equivalent parts per million (p.p.m.) of calcium carbonate irrespective of the actual salts present.

EXPERIMENTAL PROCEDURE

(*a*) *Total hardness (calcium and magnesium)*—
  (i) *Using standard soap solution*—Measure 100 ml of the water into a stoppered flask or bottle. Run in standard Wanklyn's soap solution from a burette, 1·0 ml at a time, until a lather is obtained on shaking, which covers the whole surface and persists for one minute.

If the titre on 100 ml of water is T ml,
since 1·0 ml standard soap solution $\equiv$ 1·0 mg $CaCO_3$
∴ 100 parts of water contain $0.001T$ parts of $CaCO_2$
and 1 000 000 parts of water contain $10T$ parts of $CaCO_3$
∴ Total hardness = $10T$ p.p.m. as $CaCO_3$

  (ii) *Using EDTA reagent*—Measure 100 ml of the water into a 250 ml conical flask. Add 2·0 ml of ammonia buffer solution followed by one tablet of total hardness indicator. Allow to dissolve and titrate with 0·02N-EDTA solution from a burette until the colour changes from red to blue or grey. There must be no trace of red or purple. It may be easier to see the end-point if a white porcelain evaporating dish is used in place of a flask and the contents stirred with a glass rod.

Since 1·0 ml 0·02N EDTA solution $\equiv$ 1·0 mg $CaCO_3$
Total hardness = $10T$ p.p.m. as $CaCO_3$

(*b*) *Calcium hardness*—Measure 100 ml of the water sample into a flask or evaporating dish. Add 1 ml of 4N-sodium hydroxide and one tablet of calcium hardness indicator. Titrate with 0·02N-EDTA solution until the solution becomes violet. (0·1 ml causes no further colour change).

$$\text{Calcium hardness} = 10T \text{ p.p.m. as } CaCO_3$$

(*c*) *Magnesium hardness*—This is obtained by difference.

$$\text{Magnesium hardness} = \text{total hardness} - \text{calcium hardness.}$$
$$\text{(all values expressed as p.p.m. } CaCO_3\text{).}$$

Magnesium hardness (p.p.m. $MgCO_3$) = 0·84 × magnesium hardness (p.p.m. $CaCO_3$)

(*d*) *Temporary and permanent hardness*—Titrate 100 ml of the water for total hardness with 0·02N-EDTA solution as described previously. Take another 100 ml portion of the water and boil gently for 10 min in a 250 ml flask. Cool and filter, using a funnel and filter paper, into a 250 ml conical flask. Do not wash the filter paper but dilute the filtrate with distilled water to make up the volume to about 100 ml, i.e. replace the water lost by evaporation. Titrate as for total hardness with 0·02N-EDTA solution to obtain the permanent hardness.

Temporary hardness = Total hardness − permanent hardness
(all values expressed as p.p.m. $CaCO_3$).

NOTES

1. If suitable samples of hard water are not available they can be made as follows:

(i) *Permanent hard water containing calcium and magnesium*—Dissolve A.R. calcium carbonate (0·20 g) and A.R. magnesium oxide (0·10 g) in 0·10N-hydrochloric acid (90 ml) and make up to 2·5 litres with distilled water.

(ii) *Temporary and permanent hard water*—Dissolve A.R. calcium carbonate (0·40 g) in 0·10N-hydrochloric acid (80 ml). Add A.R. sodium bicarbonate (0·34 g), make up to 2·5 litres with distilled water and shake thoroughly.

With these solutions, the experimental results obtained may be compared with the theoretical values.

(2) Method (b) for calcium hardness may be modified to determine trace amounts of calcium in materials, for example the ash from approx. 0·5 g of wool (*w* g) is dissolved in 2N-hydrochloric acid (2 ml), neutralised with 2N-sodium hydroxide, 4N-sodium hydroxide (1·0 ml) is added, the solution diluted with distilled water to about 25 ml and titrated with 0·01N-EDTA solution (*T* ml) in the presence of a calcium hardness indicator tablet.

1·0 ml 0·01N-EDTA solution ≡ 0·000 20 g calcium

$$\% \text{ calcium} = \frac{0 \cdot 02T}{w}$$

REFERENCE
1 VOGEL, A. I., *A Textbook of Quantitative Inorganic Analysis*. Longmans, London 415 (1961)

3 Estimation of iron in water

APPARATUS

Six 50 ml Nessler tubes. Spekker or EEL photoelectric absorptiometer (optional).

MATERIALS

A.R. Ammonium ferrous sulphate, potassium permanganate, potassium thiocyanate, samples of water contaminated with iron (Note 1).

THEORY

Ferric iron reacts with thiocyanate ions to give an intense red colour due to a ferric thiocyanate complex. This is used as the basis of a colorimetric method for the determination of iron.

EXPERIMENTAL PROCEDURE

(*a*) *Preparation of solutions*—Dissolve ammonium ferrous sulphate (0·702 g) in distilled water (300 ml) containing conc. sulphuric acid (2 ml) and dilute to 1000 ml in a graduated flask. Pipette 25 ml of this solution into a graduated flask and make up to 250 ml with distilled water. The final standard iron solution contains 10 microgrammes (μg) of iron per millilitre (10 p.p.m. Fe).

Prepare dilute (0·2 per cent) potassium permanganate solution by making 0·20 g of solid up to 100 ml with water (shake well to ensure that all dissolves) and 20 per cent potassium thiocyanate solution by making up 20 g of solid to 100 ml with water.

(*b*) *Determination of iron*—Measure 50 ml of the water sample into a Nessler tube. Into five other Nessler tubes run in 5, 10, 15, 20 and 25 ml respectively of the standard iron solution from a burette. Make each up to the 50 ml mark witn distilled water. The five standard solutions contain 1·0, 2·0, 3·0, 4·0 and 5·0 p.p.m. of iron respectively.

To each of the six Nessler tubes add 0·2 per cent potassium permanganate solution drop by drop until a very slight pink coloration remains after shaking. The addition may be made from a dropping bottle or burette. This is to ensure that all the iron present is in the ferric state.

To each solution, add 2N-hydrochloric acid (5·0 ml) and 20 per cent potassium thiocyanate solution (5·0 ml) and shake well. Compare the colour of the test sample with those of the standard solutions and estimate the amount of iron present (*see* Note 2). Comparison should be made soon after preparation since the colour fades on standing.

Results less liable to personal error and which avoid visual interpolation between standards are obtained with a Spekker photoelectric absorptiometer (*see* page 48) using a 1 cm cell and No. 3 filter. A calibration curve is plotted for the standard solutions and the value for the test solution is read off from this. Alternatively, an EEL absorptiometer may be used with a 1 cm cell and No. 603 filter.

NOTES

1. If the tap water does not contain sufficient iron for satisfactory analysis, samples for testing may be made by suitably diluting ammonium ferrous sulphate solution (0·702 g/l) which contains 100 p.p.m. Fe.

2. If the test sample is assessed visually and it lies between two standards, the result should be given to the nearest 0·5 p.p.m. With a photoelectric absorptiometer, the iron may be estimated to within 0·2 p.p.m. from the calibration graph. If the concentration of iron lies outside the range of the standards, the test sample should be suitably diluted.

REFERENCE
1 VOGEL, A. I., *A Textbook of Quantitative Inorganic Analysis*. Longmans, London, 785 (1961)

4 Determination of the saponification value of an oil

APPARATUS

No special apparatus is required.

MATERIALS

Approximately 0·5N-alcoholic potassium hydroxide (7 g potassium hydroxide + 250 ml of 95 per cent ethyl alcohol), 0·5N-hydrochloric acid (normality exact), phenolphthalein indicator solution, various vegetable oils.

EXPERIMENTAL PROCEDURE

Weigh a small specimen tube empty and reweigh containing about 2 ml of the oil. Insert the tube and contents into a 100 ml round-bottomed flask, add exactly 25·0 ml of the alcoholic potassium hydroxide solution from a pipette, and fit with a reflux condenser. Boil, preferably on a water bath for 1 h. Allow to cool, disconnect the flask, add 5 drops of phenolphthalein indicator and titrate the residual alkali with 0·5N hydrochloric acid. Agitate, with a circular motion, particularly near the end-point, to ensure all alkali is removed from the tube.

At the same time carry out a blank determination by refluxing 25·0 ml of the alcoholic potassium hydroxide solution for 1 h without any oil. This is to correct for any neutralisation of the alkali by carbon dioxide from the atmosphere.

THEORY AND CALCULATION OF RESULTS

(For references *see* page 15)

Vegetable and animal fats and oils consist of esters of fatty acids with glycerol. These saponify with alkali and the saponification value is defined as the number of milligrammes of potassium hydroxide required to hydrolyse 1 g of oil or fat.

$$\begin{array}{l} CH_2OCOR \\ | \\ CHOCOR \quad + \; 3KOH \; \rightarrow \\ | \\ CH_2OCOR \end{array} \qquad \begin{array}{l} CH_2OH \\ | \\ CHOH \quad + \; 3RCOOK \\ | \\ CH_2OH \end{array}$$

$$\text{Glyceride} \qquad\qquad\qquad \text{Glycerol} \qquad \text{Soap}$$

In the experimental procedure described, if the difference between the determination and the blank is x ml of $0.5N$ hydrochloric acid and w g of oil is used,

$$1.0 \text{ ml of } 0.5N \text{ HCl solution} \equiv 28 \text{ mg potassium hydroxide}$$
$$x \text{ ml of } 0.5N \text{ HCl solution} \equiv 28x \text{ mg potassium hydroxide}$$

w g of oil require $28x$ mg potassium hydroxide

$$\text{Saponification value} = \frac{28x}{w}$$

5 Determination of the iodine value of an oil

APPARATUS

No special apparatus is required.

MATERIALS

Iodine trichloride (ICl_3) and iodine *or* iodine monochloride (ICl), glacial acetic acid, carbon tetrachloride, potassium iodide, 0·10N-sodium thiosulphate (*see* page 4), various vegetable oils.

EXPERIMENTAL PROCEDURE

(*a*) *Preparation of Wij's solution* —Mix a solution of iodine trichloride (10·0 g) in glacial acetic acid (300 ml) with a solution of iodine (10·5 g) in carbon tetrachloride (300 ml) and make up to 1 litre with glacial acetic acid or make up iodine monochloride (5·0 ml) to 1 litre with glacial acetic acid. Store in the dark. (CAUTION—Iodine trichloride and monochloride should be handled with care and it is convenient for these solutions to be available to students ready prepared. Pipettes must not be filled with Wij's solution by mouth, but by means of a rubber bulb. If the latter is not available, a water pump may be used to apply gentle suction.)

(*b*) *Analysis of an oil*—Weigh 0·1 g to 2·5 g, depending on the anticipated iodine value, of the oil into a small tared tube. Place the tube and contents in a clean dry 500 ml conical flask and dissolve the oil in carbon tetrachloride (100 ml). Add Wij's solution (20·0 ml) from a pipette (*see* caution note above) and replace the stopper after moistening it with 10 per cent potassium iodide solution. Stand the flask in the dark for 30 min at 15–20°C. Add 10 per cent potassium iodide solution (15 ml) and then distilled water (100 ml), part of which is used to wash off the stopper, and finally titrate the liberated iodine with 0·10N-sodium thiosulphate. It is unnecessary to remove the tube used for weighing out the oil. Iodine will partition between the organic and aqueous phases and the end-point is reached when all the pink coloration is removed from the carbon tetrachloride layer. Sufficient time and agitation must be allowed for this to occur otherwise the titration will be taken beyond the end-point.

Carry out a parallel determination exactly as above but omitting the oil.

THEORY AND CALCULATION OF RESULTS

Vegetable oils are used widely in the textile and other industries. Chemically, they contain esters of unsaturated fatty acids and it is necessary often to quantitatively assess their degree of unsaturation. Since these oils react with halogen (X) by addition across double bonds

$$\begin{array}{c} \diagdown \quad \diagup \\ C{=}C \\ \diagup \quad \diagdown \end{array} + X_2 \rightarrow \begin{array}{c} X \quad X \\ | \quad | \\ {-}C{-}C{-} \\ | \quad | \end{array}$$

a convenient measure of their unsaturation is the iodine value which is defined as the number of grammes of iodine (or halogen expressed as iodine) which will combine with 100 g of the oil.

Since iodine itself reacts rather slowly, Wij's solution uses iodine monochloride, but in the calculation all the reacted halogen is expressed as iodine.

In the experimental procedure described, if the difference between the determination and the blank is x ml of 0·10N-sodium thiosulphate and w g of oil is used,

$$1\text{·}0 \text{ ml of } 0\text{·}10\text{N-}Na_2S_2O_3 \text{ solution} \equiv 0\text{·}0127 \text{ g iodine}$$
$$x \text{ ml of } 0\text{·}10\text{N-}Na_2S_2O_3 \text{ solution} \equiv 0\text{·}0127x \text{ g iodine}$$

$$\therefore 100 \text{ g of oil will absorb } \frac{0\text{·}0127x}{w} \times 100 \text{ g iodine}$$

$$\text{Iodine value} = \frac{1\text{·}27x}{w}$$

By consulting a table of constants for oils, fats and waxes (*see Handbook of Chemistry and Physics* [3]) it is possible to calculate a suitable weight for analysis, for example 0·3 g for olive, castor and rape oils and 0·1–0·15 g for linseed.

REFERENCES
1 FIESER, L. F. and FIESER, M., *Organic Chemistry*. Reinhold, New York, 399 (1956)
2 HILDITCH, T. P., *The Chemical Constitution of Natural Fats*. Chapman & Hall, London (1947)
3 *Handbook of Chemistry and Physics*. Chemical Rubber Publishing Company, Cleveland, U.S.A.

6 Use of the pH meter

APPARATUS

Any pH meter produced by a well-known maker is satisfactory (*see* Vogel [1] page 923), glass electrode, standard half-calomel cell, mechanical shaker.

MATERIALS

A.R. Potassium hydrogen phthalate, A.R. sodium borate, $Na_2B_4O_7 \cdot 10H_2O$, miscellaneous organic and inorganic acids and salts, B.D.H. 'Universal' indicator, samples of wool (loose, yarn, fabric etc.).

THEORY

A convenient measure of the acidity or alkalinity of a solution is the negative logarithm, to base 10, of its hydrogen ion concentration in g-ions per litre. This value is known as the pH and hence $pH = -\log_{10}[H^+]$.

Since for pure water, $[H^+] = [OH^-] = 10^{-7}$, an exactly neutral solution has a pH value of 7 (quantities in square brackets indicate concentrations in g-ions per litre). Numerical values below 7 indicate that the solution is acid and above 7 that the solution is alkaline. For most practical purposes, pH values lie within the range 0–14 (normal acid to normal alkali). For a more extended account of the concept of pH (*see* Vogel [1] page 25).

Since many textile fibres are susceptible to degradation in acid or alkaline solution, it is common practice to specify the pH of solutions used in their wet processing, for example scouring, dyeing, bleaching. Although indicators are widely used for this purpose, if the solution is coloured or a more accurate result is required, it is usual to employ some type of pH meter. Briefly, a suitable electrode is immersed in the solution of unknown pH and the resultant half-cell is coupled with a standard half-cell (generally calomel). Instruments which measure the e.m.f. produced are essentially valve voltmeters and they may be either of the direct reading type or they may employ a null-point indication.

EXPERIMENTAL PROCEDURE

(*a*) *Operation of the pH meter* —In addition to following the manufacturer's instructions supplied with the meter, attention should be paid to the following points

Suitable standard buffer solutions for calibrating the meter are 0.05M-potassium hydrogen phthalate (10.211 g/l) and 0.05M-sodium borate (19.072 g $Na_2B_4O_7 \cdot 10H_2O$ per litre) of pH 4.00 and 9.21 respectively at 20 C.

Generally, the electrode system should be washed by repeated immersion in portions of distilled water to remove previous standardising or test solutions until the indicated pH value shows no change. Then determine the pH of successive portions of the test solution until a constant value is obtained. This is particularly important if the test solution is unbuffered. If a number of buffered solutions of similar pH are being tested, however, one wash with the test solution followed by the determination on a fresh portion will usually suffice.

The effect of temperature on the pH of a solution is appreciable and it is very possible that the sample removed, for example from a dyebath, will be at a different temperature from the solution used for calibration. If this occurs, the solution should either be cooled or the temperature control on the meter adjusted.

The glass electrode is fragile and mechanical shock must be avoided. It should be stored always in distilled water and not left longer than necessary in solutions of pH greater than 9. It will give erroneous results in solutions of pH above 10 and for solutions of very high pH special electrodes should be used.

(*b*) *Determination of the pH value of water extracts of wool*—Thoroughly clean a hard-glass, for example 'Pyrex' 250 ml conical flask and wash out well with distilled water. Add the wool (2·0 g) and distilled water (100 ml) The pH of the latter should lie between 5·0 and 6·5. Shake vigorously by hand to thoroughly wet out the wool and then shake mechanically for 1 h. Wash out the pH cell with distilled water as described under (a) and determine the pH of successive portions of the solution decanted from the wool until a constant value is obtained, which is recorded as the pH of the extract.

To 10·0 ml of the extract, add 0·2 ml of B.D.H. 'Universal' indicator solution. Note the colour and determine the pH from the colour code provided. Compare the two results.

For additional practice, determine the pH by the meter and 'Universal' indicator of some of the solutions given by Vogel [1] pp. 1161 and 1162 and compare the values obtained with those given.

It should be noted that for most practical purposes, pH values need not be determined with an accuracy greater than 0·1 unit and this should be attained by any pH meter. 'Universal' indicator gives values within only 0·5 of a unit but 'Narrow-range' indicator papers (for example Johnson's) can reduce this margin.

REFERENCE
1 VOGEL, A. I., *A Textbook of Quantitative Inorganic Analysis*. Longmans, London (1961)

7 Use of pH indicators in acid-base titrations

APPARATUS

pH meter, glass electrode, standard half-calomel cell, electric stirrer.

MATERIALS

Ampoules of hydrochloric acid, acetic acid, sodium and ammonium hydroxides for making exactly 0.10N-solutions. Various indicators including methyl orange, methyl red, phenolphthalein and thymolphthalein.

THEORY

A neutralization or acid-base indicator is a substance which possesses different colours according to the hydrogen-ion concentration of its solution. The colour change is not sharp and usually takes place over a range of two units of pH.

The object of titrating a solution of an alkali with a standard solution of an acid is to determine the amount of acid which is exactly chemically equivalent to the alkali present. This point is the equivalence point or theoretical end-point. It must be appreciated that at this point the resulting solution may not be at pH 7. Thus, if stoichiometrical quantities of a strong base NaOH and a weak acid HA are mixed, at the theoretical end-point the salt NaA will be hydrolysed.

$$\text{NaA} + \text{H}_2\text{O} \rightleftharpoons \text{Na}^+ + \text{OH}^- + \text{HA}$$

and the solution may have a pH as high as 9 or 10.

C

Conversely, if a weak base is titrated against a strong acid, the solution will be acid at the theoretical end-point. Clearly, for such titrations the end-point obtained in practice will only correspond to the theoretical end-point if an indicator changing colour at the point of equivalence is used.

If a known volume of an acid is taken, and this is titrated against an alkali of the same normality, by immersing an electrode in the solution while the titration proceeds, a neutralisation curve may be plotted of pH against volume of alkali added (*see* Figure 1). From

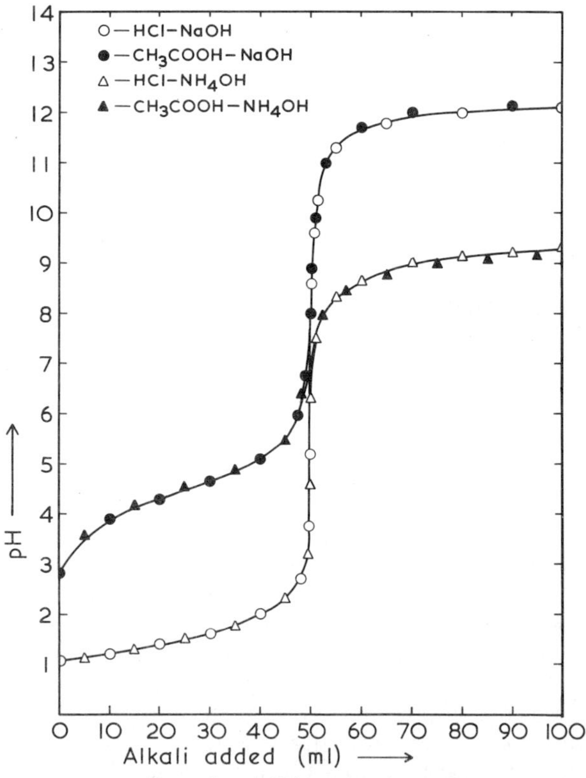

Figure 1.   Acid-base titration curves

this curve the pH at the equivalence point may be obtained and a suitable indicator selected for subsequent visual titrations. Since indicators change colour over a range of pH values, unless there is an abrupt change in pH of at least 1 or 2 units at the equivalence point, it may be difficult to select a satisfactory indicator for a particular titration.

EXPERIMENTAL PROCEDURE

Measure accurately 50·0 ml of 0·10N-hydrochloric acid into a 250 ml beaker. Insert an electric stirrer and glass-calomel electrode system into the beaker taking care not to knock or scratch the glass electrode. Couple the electrodes to a pH meter and at intervals run in 0·10N-sodium hydroxide from a burette (*see* Figure 2). Stir well after

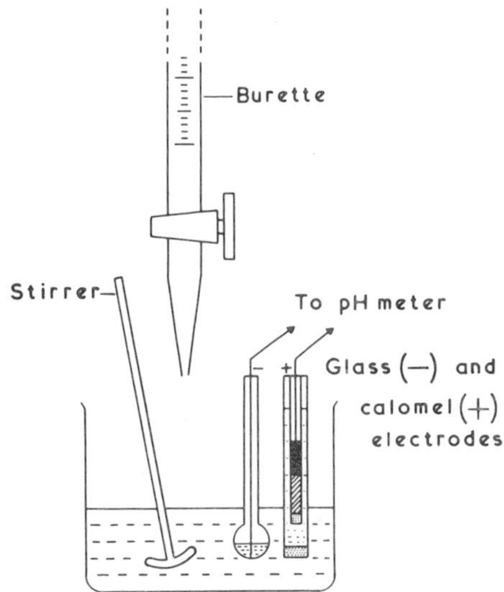

Figure 2.   Apparatus for potentiometric titrations

each addition and measure the pH (*see* page 16). Add the alkali in 5 ml portions up to 40 ml, then in 1 ml portions up to 49 ml and then in 0·10 ml portions up to 50 ml. Use corresponding quantities in reverse order from 50 to 100 ml. Plot the pH value against the volume of alkali added (neutralisation curve (a)).

In a similar manner construct the following neutralisation curves (first named reagent in burette).

(b) 0·10N-sodium hydroxide *v.* 0·10N-acetic acid.
(c) 0·10N-ammonium hydroxide *v.* 0·10N-hydrochloric acid.
(d) 0·10N-ammonium hydroxide *v.* 0·10N-acetic acid.

From these curves determine the pH of the solution at the theoretical end-point and from the following select suitable indicators for each titration (pH range in parentheses), methyl orange (3·1–4·4), methyl red (4·2–6·3), phenolphthalein (8·3–10·0) and thymolphthalein (9·3–10·5).

Finally, carry out titrations (a) to (d) as in normal volumetric analysis using the indicator selected and comment on their suitability to give an accurate visible indication of the equivalence point of the titration system.

Solutions of the indicators required are prepared as follows:

Methyl Orange   Dissolve 0·5 g of the free acid in 1 litre of water.
Methyl Red   Dissolve 1·0 g of the free acid in 1 litre of hot water.
Phenolphthalein   Dissolve 5 g of solid in 500 ml alcohol and add 500 ml water with stirring.
Thymolphthalein   Dissolve 0·4 g of solid in 600 ml alcohol and add 400 ml water with stirring.

REFERENCE
1 VOGEL, A. I., *A Textbook of Quantitative Inorganic Analysis*. Longmans, London, 51 (1961)

8 Determination of the nitrogen content of an organic substance by the Kjeldahl method

APPARATUS

Quick-fit macro-Kjeldahl assembly, electric heating mantle.

MATERIALS

Potassium sulphate (A.R.), copper sulphate, boric acid, sodium hydroxide, 0·10N-sulphuric acid, methyl red, acetanilide (analytical standard).

EXPERIMENTAL PROCEDURE

Weigh accurately a quantity of the material containing approx. 30 mg of nitrogen. It should be borne in mind that most substances to be analysed are not completely unknown and that this method is particularly useful for proteins, all of which contain approx. 16 per cent of nitrogen. Place in a dry long-necked Kjeldahl flask and add potassium sulphate (2 g), copper sulphate (0·2 g) and conc. sulphuric acid A.R. (5 ml).

Insert a small funnel into the neck of the flask, place in a slightly inclined position in a fume cupboard and heat over a small flame nearly to b.p. until the initial decomposition is complete. This may be regarded as having taken place when active evolution of sulphur dioxide has ceased and the liquid has acquired a uniform brown colour with no lumps of solid remaining. This will normally take about 30 min. Now heat to gentle boiling until the solution becomes

clear or a pale green colour ($1\frac{1}{2}$–3 h) and continue gentle boiling for
a further 30 min. It is essential that at the end of the digestion no
partially decomposed material adheres to the sides of the flask.

Allow the solution to cool and dilute to approx. 200 ml by the
careful addition of distilled water. Add a suitable anti-bumping
agent (previously ignited porous pot) and set up with the rest of the

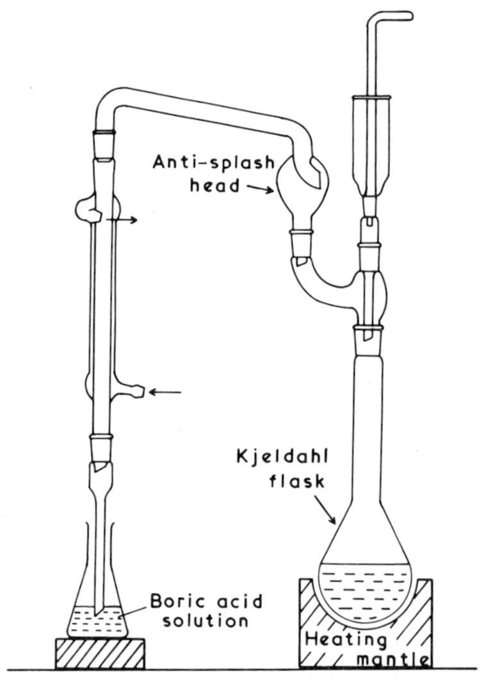

Figure 3. Kjeldahl apparatus for nitrogen determination

assembly for distillation (Figure 3). Ensure that the ground-glass
joints are greased and air-tight and lag the horizontal portion of the
assembly between the splash-head and the condenser. Place about
40 ml of saturated (approx. 4 per cent) boric acid in the receiving
flask and ensure that the tip of the delivery tube dips just below the
surface of the absorption solution. When set up in this manner, no
liquid can be sucked into the distillation flask when cooling of the
latter takes place. The delivery tube should be adjusted during the
course of the distillation so that it is always in the correct position.

Neutralise the acid in the distillation flask by adding sodium hydroxide solution (20 g sodium hydroxide $+50$ ml water) through the funnel provided, taking care that no ammonia can escape. Bring the flask to the boil using an electric heating mantle. This should ensure even heating and prevent breakage of the flask through thermal shock. Continue the distillation for 30–45 min, and then wash the condenser down with distilled water. Titrate the absorption solution with 0·10 N-sulphuric acid using methyl red or a mixture of methyl red and bromocresol green as indicator.

Carry out a blank determination using exactly the same quantities of reagents under the same conditions and correct the actual determination for any nitrogen found in the blank.

As a check on the accuracy of the method, determine the nitrogen content of pure acetanilide, calculating the necessary quantity from its formula. Finally determine the nitrogen content of wool (0·2 g), silk (0·2 g), nylon (0·2 g) or a urea-formaldehyde resin (0·1 g).

THEORY AND CALCULATION OF RESULTS

The majority of organic nitrogen compounds, in which the nitrogen exists in a non-oxidised form, when heated with concentrated sulphuric acid are decomposed with the quantitative formation of ammonium sulphate. To assist this decomposition potassium sulphate is added to raise the b.p. of the solution and copper sulphate acts as a catalyst. Other suitable catalysts are mercury and selenium.

Analysis by this method of nitro, nitroso and azo compounds is not recommended, but it is very satisfactory for amines, amides, proteins etc. The ammonium sulphate is decomposed by alkali and the liberated ammonia titrated with acid.

$$\equiv\!N + H_2SO_4 \rightarrow (NH_4)HSO_4$$

$$(NH_4)HSO_4 + 2NaOH \rightarrow Na_2SO_4 + NH_3 + 2H_2O$$

$$1 \text{ litre N-}H_2SO_4 \text{ solution} \equiv 17\,g\,NH_3 \equiv 14\,g\,N$$

$$1\!\cdot\!0 \text{ ml } 0\!\cdot\!10\text{N-}H_2SO_4 \text{ solution} \equiv 0\!\cdot\!0014\,g\,N$$

If w g of substance requires a titre of v ml 0·1N-H_2SO_4 solution

$$\text{Percentage N} = \frac{0\!\cdot\!14v}{w}$$

Boric acid is a weak acid which 'fixes' the ammonia as ammonium borate which titrates with the strong acid.

9 Paper chromatography—principles and techniques

APPARATUS

Shandon 'Unikit' No. 1*, hair-dryer.

MATERIALS

Whatman No. 1 chromatography paper, various coloured inks and indicators (these are contained in Part M of the 'Unikit'), n-butanol, ethanol (industrial methylated spirits), glacial acetic acid.

DISCUSSION

Paper chromatography is the technique of separation and identification of chemical substances by a moving solvent on sheets or strips of filter paper. The distance travelled by a substance in a given time is the resultant of various forces and depends chiefly on the absorption of the substance by the cellulose and its partition or distribution between the solvent phase flowing over, and the aq. phase normally bound to the cellulose. The rates of migration of substances often come close to the theory for a simple liquid–liquid distribution but sometimes adsorption effects appear to operate.

It is usual to define migration rates (R_F values) by the ratio

$$R_F = \frac{\text{Distance substance has moved from origin}}{\text{Distance solvent front has moved from origin}}$$

* Suitable apparatus for carrying out paper and thin layer chromatography and electrophoresis is contained in Shandon 'Unikits' obtainable from Shandon Scientific Company Ltd., 65 Pound Lane, London, N.W.10.

the solvent front being the furthest part reached by the advancing solvent on the paper. The R_F value may be expressed as a decimal or ($\times 100$) as a percentage. To increase the resolution of spots near the origin, it may be necessary to run the solvent front off the paper in which case it is not possible to measure the R_F value directly, but by reference to the distance travelled by a known substance (x)

$$R_x = \frac{\text{Distance substance has moved from origin}}{\text{Distance substance } x \text{ has moved from origin}}$$

If the substances being separated are coloured, for example dyes or DNP derivatives of amino acids, the spots are located visually. Some substances, for example DNS derivatives of amino acids may be located by their fluorescence under ultra-violet light, while others require development to a coloured substance with a suitable reagent, for example ninhydrin for amino acids.

If the resolution of a mixture of substances is incomplete when the method (one-way chromatography) outlined is used, a better separation may be achieved by running the chromatogram for a second time in a different solvent at right angles to the first direction (two-way chromatography).

EXPERIMENTAL PROCEDURE

(*a*) *One-way ascending chromatography*—Use a 25 cm square of Whatman No. 1 paper and draw a pencil (not ball-point) line across the paper 2·5 cm from one edge. Beginning 3 cm from one side, make pencil marks at 2 cm intervals along the line. Since each mark is to serve as an origin for different samples, they must be identified with a symbol in pencil (*see* Figure 4a).

Apply the test solution to the paper using the 4 mm diameter platinum-iridium wire loop supplied with the kit, to obtain a spot not greater than 5 mm in diameter. Alternatively, in later work, use a fine capillary tube to give a smaller spot. Wash the wire loop between solutions to avoid cross-contamination. Allow the spots to dry.

As a preliminary experiment, the following indicators and dye solutions supplied with the 'Unikit' may be examined—Congo red, phenol red, bromophenol blue, mixtures of indicators, Universal indicator, fluorescein, tartrazine, malachite green and a mixture of dyes.

Prepare 2N-ammonium hydroxide by diluting 0·880 ammonia solution (135 ml) to 1 litre with water and make a mixture of n-butanol–ethanol–2N-ammonium hydroxide (60:20:20 v/v). Place approximately 50 ml of this solvent in the bottom of the tank and cover with the lid to prevent evaporation.

Bring two edges of the paper together to form a cylinder with the starting line at the base. Fasten the edges together with the tongued plastic clips to keep the edges apart and arrange symmetrically to form a stable cylinder. Lower the paper cylinder into the tank

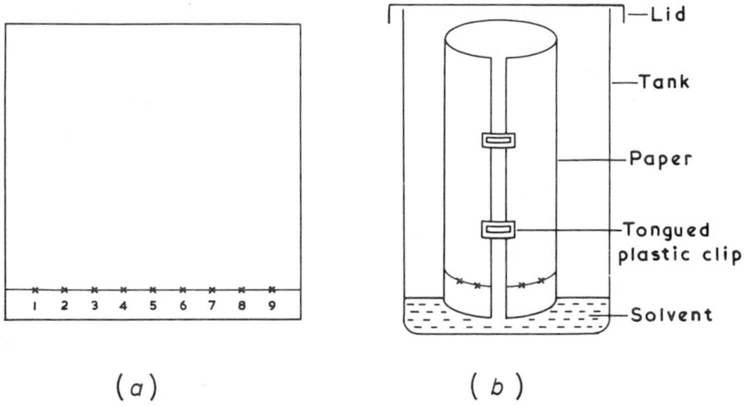

Figure 4.   One-way ascending paper chromatography

containing solvent so that it is standing freely on the bottom and not touching the sides (*see* Figure 4b). The surface of the solvent should be well below the origin line on the paper. Place the lid in position with the two holes in it closed by polythene caps. Allow to run for at least 2 h or preferably all day. Since the components are coloured, the chromatographic development can be followed visually. Remove the paper, mark the solvent front with pencil, hang in a fume cupboard by stainless steel clips and dry with a hair-dryer.

Carry out a similar experiment using as solvent n-butanol–acetic acid–water (60:15:25 v/v) and also separate the inks provided with the 'Unikit' (brown, Royal blue, scarlet, green and black) into their components, using both solvents.

Record the R_F values of all the components isolated for each solvent, measuring from the centres of the spots.

(b) *One-way descending chromatography*—Take a strip of Whatman

No. 1 paper 30 × 10 cm and starting from one end draw pencil
lines 1·5, 1·5, 2·0, 1·5 and 2·0 cm apart (Figure 5a.) The paper is
creased and folded sharply at the first four lines, whilst the origins
are marked on the fifth. Place the paper with the spots on the origins

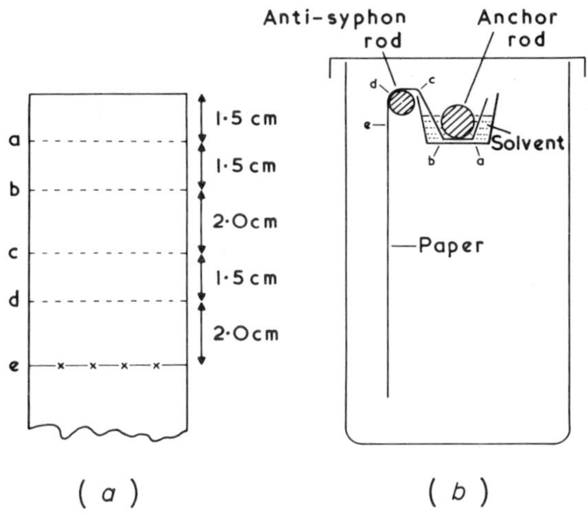

Figure 5.   One-way descending paper chromatography

in position in the plastic trough at the top of the 'Unikit' tank with
the anti-syphon and anchor rods as shown in Figure 5b. Two papers
can be used hanging down on opposite sides of the trough. Pour
the solvent into the trough after the papers are in position, taking
care to avoid splashing. Close the tank with the lid and allow the
solvent to run. When a satisfactory separation has been achieved,
first remove the anchor rod and then the paper together with the
anti-syphon rod, taking care that drops of liquid do not fall on the
chromatogram. Invert the paper, dry as in (a) and record the
R_F values.

REFERENCES
1 SMITH, I. and FEINBERG, J. G., *Paper and Thin layer Chromatography and Electro-
    phoresis*. Shandon, London (1965) .
2 BAILEY, J. L., *Techniques in Protein Chemistry*. Elsevier, Amsterdam **p. 1 (1967)**

10 Separation of amino acids by paper chromatography

APPARATUS

Shandon 'Unikit' No. 1, hair-dryer, Shandon laboratory spray gun (optional).

MATERIALS

All, or a selection of the following amino acids—alanine, arginine, aspartic acid, cysteic acid*, cystine, glutamic acid, glycine, histidine, hydroxyproline*, lanthionine*, isoleucine, leucine, lysine, methionine, phenylalanine, proline, serine, threonine, tryptophan, tyrosine, valine, mixtures of amino acids (a mixture of the following separates well under the conditions given—alanine, aspartic acid, cystine, glycine, leucine, proline, tyrosine and valine).

Ninhydrin, n-butanol, isopropanol, acetone, acetic acid, phenol, Whatman No. 1 chromatography paper, purified wool and silk.

EXPERIMENTAL PROCEDURE

Prepare the following:

(i) Approximately 0·01M solutions of amino acids by dissolving 0·01 g in a mixture of water (9 ml) and isopropanol (1 ml). Mixtures should be approximately 0·01M with respect to each amino acid. If solution is not complete, for example cystine and tyrosine, add a drop of conc. hydrochloric acid.

* The amino acids so marked are not normally found in silk and wool and may be omitted.

(ii) Hydrolyse wool or silk (0·1 g) with 5N-hydrochloric acid (2 ml) at 110°C in a sealed tube overnight, as described on page ix, and take down to dryness in a vacuum desiccator over solid sodium hydroxide. Dissolve the silk or the wool hydrolysate in solvent (10 ml) as described under (i).

(iii) Chromatography solvents consisting of (a) n-butanol (60 ml), acetic acid (15 ml) and water (25 ml) and (b) phenol (50 g) and water (12·5 ml). The latter solvent is toxic and caustic.

(iv) A developing reagent consisting of ninhydrin (0·2 g) dissolved in acetone (100 ml).

(*a*) *One-way ascending chromatography*—Prepare a 25 cm square sheet of Whatman No. 1 paper for ascending chromatography as described on page 27. It is important that hands are washed thoroughly before the paper is touched as amino acids are present in perspiration. The paper must be handled at the edges only, since finger marks show up as confusing spots after spraying with the very sensitive ninhydrin locating agent. The drop of solution applied to the paper contains only about 5 μg (1 μg = 1 microgramme = 10^{-6}g) of any individual amino acid.

Lay the square of paper on another sheet of clean paper so that the origin line hangs over the edge of the bench. Apply the solutions to the paper using a piece of capillary (melting point) tube, washing the tube thoroughly each time to avoid cross-contamination. Practice on a piece of filter paper first to obtain spots not greater than 5 mm diameter. Since the origin spots must be at least 2 cm apart, two papers will be required to run all the amino acids suggested as well as the two hydrolysates in each solvent.

Dry the spots with a hair-dryer and run the chromatogram in n-butanol–acetic acid–water solvent for 5–8 h. Remove the chromatogram from the tank, mark the solvent front with pencil, open out and suspend with stainless steel clips from a glass rod. Dry with a hair-dryer in a fume cupboard. Spray the paper uniformly with ninhydrin solution using either the atomizer supplied with the 'Unikit' or a Shandon laboratory spray gun. Allow the paper to stand for about 1 h at room temperature for colour development or the paper may be heated in an oven at about 105°C for a few minutes. The former tends to produce a cleaner chromatogram. Since the spots fade, for a permanent record their positions should be marked with pencil or ink or a photostat copy may be made. Measure the R_F values (*see* page 26) for each of the known amino acids and attempt to determine those present in the mixture and in silk and wool. Resolution of the latter chromatogram may not be complete in one direction, but estimate how many different amino acids are probably present.

Repeat the chromatograms using phenol–water solvent and air-dry overnight or longer in a fume-cupboard prior to spraying.

(*b*) *Two-way ascending chromatography*—Use a 25 cm square sheet of paper and mark two datum lines 3 cm from the edge intersecting at the origin O (*see* Figure 6). Place three drops of an unknown

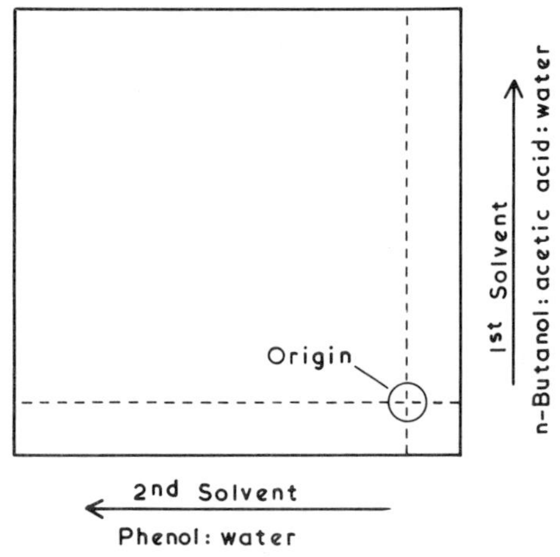

Figure 6.   Two-way ascending paper chromatography

amino acid mixture on the origin, allowing to dry between each drop. Use 50 ml of n-butanol–acetic acid–water (60:15:25 v/v) as the first solvent, running for 5–10 h. Mark the position reached by the solvent front and dry but do *not* spray with ninhydrin.

Reform the paper into another cylinder with the now partially separated amino acids forming a circle round the base of the cylinder. Allow to run for 5–8 h in phenol–water as the second solvent. Remove and air-dry in a fume cupboard overnight before spraying with ninhydrin solution to locate the component amino acids.

Repeat using a wool or silk hydrolysate and attempt to identify as many spots as possible using the previously determined R_F values in the two solvents.

REFERENCES
1 SMITH, I. and FEINBERG, J. G., *Paper and Thin layer Chromatography and Electrophoresis*. Shandon, London (1965)
2 BAILEY, J. L., *Techniques in Protein Chemistry*. Elsevier, Amsterdam, p. 1 (1967)

11 Thin layer chromatography

APPARATUS

Shandon 'Unikit' No. 2 for thin-layer chromatography

MATERIALS

Kieselgel G silica gel (Merck), micro-granular cellulose powder (Whatman CC41), range of coloured inks (supplied with kit) and amino acids (*see* page 30), organic solvents (*see* page 30).

DISCUSSION

The technique of thin-layer chromatography (TLC) has grown mainly from the pioneering work of Stahl published in 1958. The absorbing material is deposited as a thin layer about 0·25 mm thick on a plate of glass or other inert material, by spreading the absorbent on the plate as a slurry with water or other medium and then allowing to dry. Common absorbents used are silica gel and alumina, and the first applications were in the lipid field for the separation of closely related fatty acids for which paper chromatography is not satisfactory. Other absorbents such as cellulose powder can be used for the separation of hydrophilic substances such as amino acids and peptides and the separation is then similar to paper chromatography.

The advantages of thin-layer chromatography are that, in general, the spots are more compact, resolution is better and the time for separation is much shorter than is required in paper chromatography.

Extemely small quantities of sample can be used and the layer can be easily scraped off and eluted with solvent for recovery of the content of the spot or band.

As in paper chromatography, TLC plates may be run in one direction (ascending) or in two directions and if the substances being separated are not coloured the spots must be located with a suitable reagent.

EXPERIMENTAL PROCEDURE

(*a*) *Preparation of TLC Plates*—The thin layer plates (10 cm × 10 cm) or slides (10 cm × 2·5 cm) are prepared using the Shandon 'Unikit' spreader and spreading jig. As the plaster of Paris binder present in the silica gel sets in about 4 min after mixing with water, correct timing is necessary to ensure the preparation of good plates.

Weigh the Kieselgel G (10 g) into a dry 100 ml conical flask. Add distilled water (20 ml) all at once, stopper and shake thoroughly for 75 s. Pour the slurry into the spreader and draw smoothly and continuously over the clean glass plates or slides in the spreading jig taking about 5 s. Allow the plates to remain undisturbed in the jig for 5–10 min to set. Then, handling only by the edges, place in the plate holder and allow to air-dry overnight. If 'activated' plates of greater absorptive properties are required, heat the coated plates in an oven at 160–170°C for 1 h, cool and store in a desiccator over phosphorus pentoxide.

Cellulose plates are prepared in a similar manner using Whatman CC41 micro-granular cellulose powder (8 g) in water (18 ml).

(*b*) *Application of samples*—For one-way chromatograms, apply the test solutions as spots from a micropipette or capillary tube at a distance of 1·5 cm from one edge of the plate and 1·0–1·5 cm apart using the spotting guide provided with the 'Unikit'. As the thin layer cannot be marked with pencil, make small scratch marks at the edges of the plate to indicate the origin line and keep a separate plan of the samples applied and their positions. Although, in general, the spots should be made as small as possible, if the solution is very dilute, repeated applications should be made allowing the spot to dry between applications. For two-way chromatograms, the spot is applied at an origin 1·5 cm up and in from one corner of the plate.

(*c*) *Development of chromatograms*—Slide the prepared plate into

the grooves in the 'Unikit' TLC 'Chromatank' containing 25 ml
of solvent, close with the lid and allow to run until the solvent front
reaches the top of the plate (approx. 1 h). Two plates can be run
at the same time. If slides (10 cm × 2·5 cm) or half-plates (10 cm ×
5 cm) are used, they are allowed to rest nearly vertically against the
sides of the tank.

The plate is removed from the tank and allowed to dry after
blotting the lower edge on filter paper. A hair-dryer or oven may be
used to assist the drying. If the components are not coloured, they
are located with a suitable spray such as ninhydrin for amino acids.

When using volatile solvents, for example benzene, chloroform,
etc., it is advantageous to place a lining paper inside the tank to
become saturated with solvent and maintain equilibrium conditions
within the tank. This can be cut from a piece of filter paper large
enough to cover the two wide walls and base of the tank.

The following may be resolved into their components by thin-
layer chromatography.

(i) The five inks provided with the 'Unikit', using Kieselgel on the
plate and either n-butanol–ethanol–2N-ammonia (60:20:20 by
volume) or n-butanol–acetic acid–water (60:15:25 by volume) as
solvent.

(ii) A mixture of up to 10 amino acids on Kieselgel or Whatman
CC41 cellulose powder and n-butanol–acetic acid–water as solvent.

(iii) The same mixture as (ii) using phenol–water (80:20 by weight)
as the solvent in the second direction (CAUTION).

REFERENCES
1 SMITH, I. and FEINBERG, J. G., *Paper and Thin-Layer Chromatography and Electrophoresis*. Shandon, London (1965)
2 BAILEY, J. L., *Techniques in Protein Chemistry*. Elsevier, Amsterdam, 36 (1967)

D

12 Chromatography on ion-exchange paper

APPARATUS

Shandon 'Unikit' No. 1.

MATERIALS

Whatman ion-exchange papers DE81 and P81, A.R. sodium chloride, sodium acetate, acetic acid, n-butanol, phenol, ninhydrin locating reagent (*see* page 31) range of amino acids, wool hydrolysate (*see* page 31).

DISCUSSION

In Whatman ion-exchange papers DE81 and P81, the primary hydroxyl group attached to carbon atom 6 of the cellulose molecule has been modified. DE81 is diethylaminoethyl cellulose (DEAE cellulose) and P81 is cellulose phosphate.

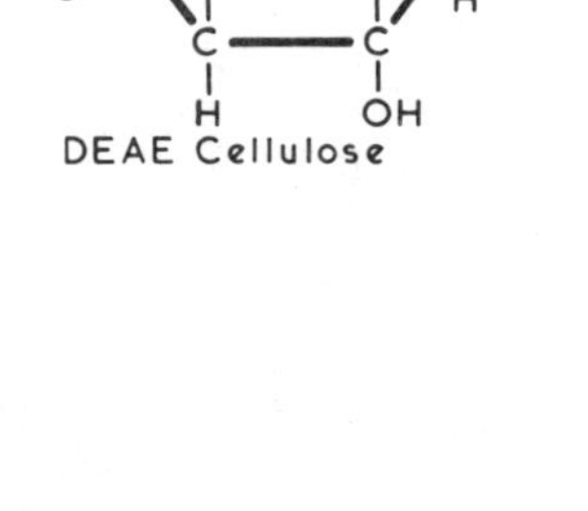

DEAE Cellulose

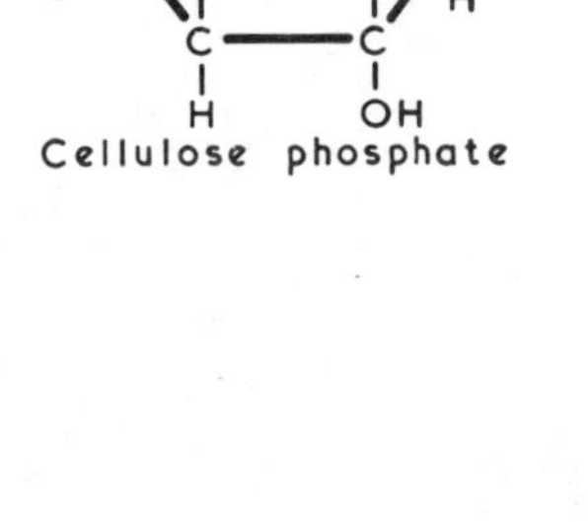

Cellulose phosphate

In the case of DEAE cellulose, the tertiary amine group is basic and reacts with amino acids as follows:

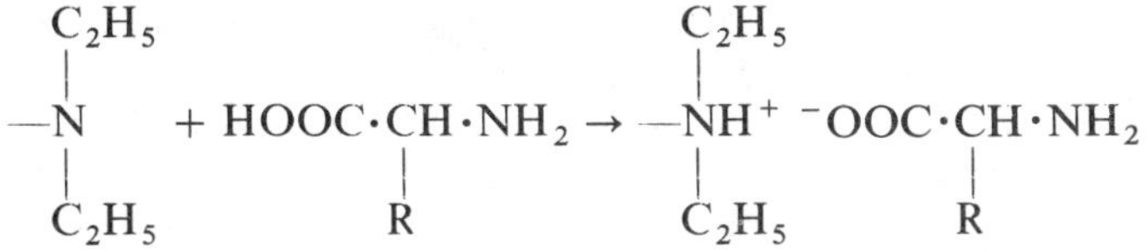

$$\underset{\underset{C_2H_5}{|}}{\overset{\overset{C_2H_5}{|}}{-N}} + HOOC\cdot \underset{R}{\overset{|}{CH}}\cdot NH_2 \rightarrow -NH^{+}\ {}^{-}OOC\cdot \underset{R}{\overset{|}{CH}}\cdot NH_2$$

When treated with a solution of an electrolyte, ion exchange occurs:

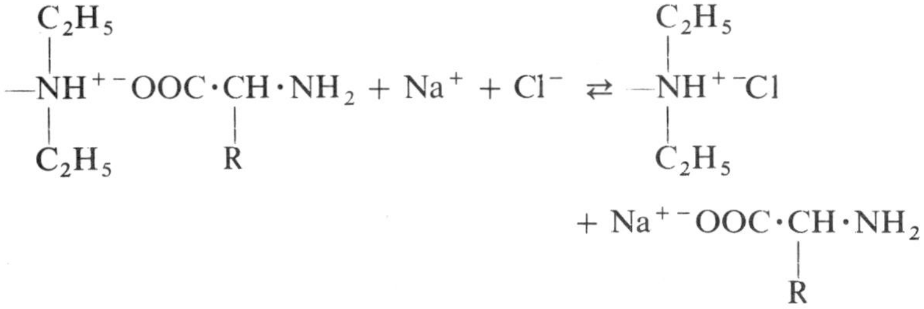

$$\underset{\underset{C_2H_5}{|}}{\overset{\overset{C_2H_5}{|}}{-NH^{+}}}{}^{-}OOC\cdot \underset{R}{\overset{|}{CH}}\cdot NH_2 + Na^{+} + Cl^{-} \rightleftharpoons \underset{\underset{C_2H_5}{|}}{\overset{\overset{C_2H_5}{|}}{-NH^{+}}}{}^{-}Cl$$

$$+ Na^{+}\ {}^{-}OOC\cdot \underset{R}{\overset{|}{CH}}\cdot NH_2$$

DEAE cellulose, therefore, acts as an anion (negative ion or acid radical) exchanger. Cellulose phosphate, on the other hand, which is acidic is a bifunctional cation exchanger.

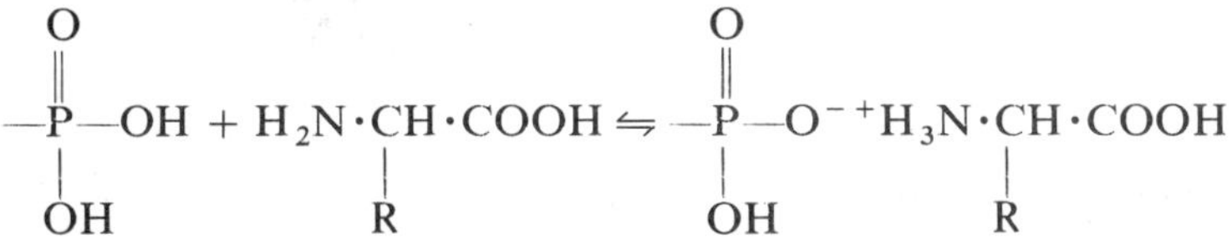

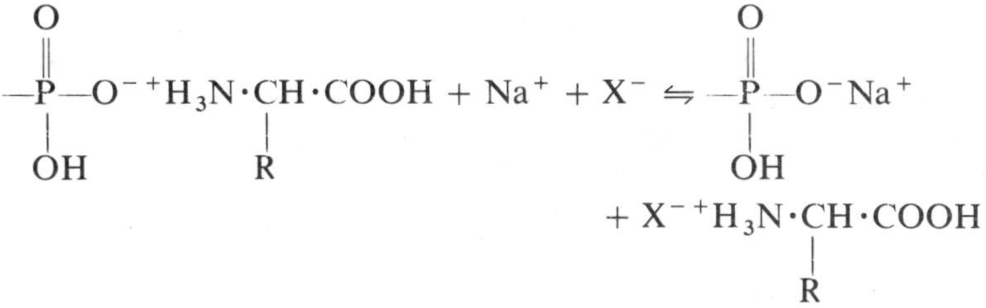

$$\underset{\underset{OH}{|}}{\overset{\overset{O}{\|}}{-P}}-OH + H_2N\cdot \underset{R}{\overset{|}{CH}}\cdot COOH \rightleftharpoons \underset{\underset{OH}{|}}{\overset{\overset{O}{\|}}{-P}}-O^{-}\ {}^{+}H_3N\cdot \underset{R}{\overset{|}{CH}}\cdot COOH$$

$$\underset{\underset{OH}{|}}{\overset{\overset{O}{\|}}{-P}}-O^{-}\ {}^{+}H_3N\cdot \underset{R}{\overset{|}{CH}}\cdot COOH + Na^{+} + X^{-} \rightleftharpoons \underset{\underset{OH}{|}}{\overset{\overset{O}{\|}}{-P}}-O^{-}Na^{+}$$

$$+ X^{-}\ {}^{+}H_3N\cdot \underset{R}{\overset{|}{CH}}\cdot COOH$$

For simplicity, the above equations are shown as if the cellulose phosphate were monofunctional.

An advantage of ion-exchange paper over ordinary paper is that by varying the solvent, chromatograms may be run either with the ion-exchange mechanism operating or a conventional partition chromatogram may be produced. In two dimensional chromatography using ordinary paper, even the most dissimilar pair of 'partition' solvents tend to produce spots lying along a roughly diagonal band on the paper. If ion-exchange paper is used, functioning as an ion-exchanger in one direction and normal paper in the other, a much better spread of randomly spaced spots is obtained. Furthermore, the ionic absorption tends to prevent spreading of the spots. This applies, also, to the original spot, because although the applied test solution appears to spread, the dissolved ionic substances are absorbed in a small area at the point of application.

EXPERIMENTAL PROCEDURE

(*a*) *Separations on DEAE paper* —Prepare 0·001M aq. sodium chloride by making up sodium chloride (5·85 g) to 1 litre with water and diluting 10·0 ml of the resulting solution to 1 litre. Using this solution as solvent, determine the R_F values of the amino acids as in normal partition paper chromatography (page 31). Spray the paper with 1 per cent acetic acid solution prior to locating the spots with ninhydrin to neutralise the alkalinity of the paper and provide the correct pH conditions for colour development.

Using a fresh sheet of DE81 paper, repeat the separation using as solvent n-butanol–acetic acid–water (60:15:25 by volume) which suppresses the ion-exchange characteristics of the paper and the separation is achieved by conventional partition chromatography.

Finally, carry out a two-way separation of a mixture of amino acids, for example the mixture given on page 30) or a wool hydrolysate (page 31) using the two solvents in the order given.

(*b*) *Separation on cellulose phosphate paper*—Prepare 0·02M-sodium acetate by making up 1·6 g of anhydrous sodium acetate or 2·7 g of the hydrated salt (CH_3COONa, $3H_2O$) to 1 litre with water. Separate the amino acids as described for DE81 paper using 0·02M–sodium acetate adjusted to pH 4·7 with acetic acid. For two-way chromatography, use phenol–water (80:20 w/w) as the second solvent (*see* page 31).

REFERENCES
1 KNIGHT, C. S., *J. Chromat.* **8**, 205 (1962)
2 KNIGHT, C. S., *Nature, Lond.* **184**, 1486 (1959)
3 KNIGHT, C. S., *Protides of the Biological Fluids,* Elsevier, Amsterdam, **10**, 318 (1963)

13 Separation of amino acids by paper electrophoresis

APPARATUS

Shandon 'Unikit' No. 1

MATERIALS

Cysteic acid, aspartic acid, alanine, lysine, acetic acid, pyridine, ninhydrin reagent, Whatman No. 1 paper.

THEORY

When the substances in a mixture are ionized or when only some ionise, a degree of separation can be obtained by subjecting the mixture dissolved in a conducting buffer solution to an electric field. It is an incomplete form of electrolysis in which the substances are stopped along their path of migration to the attracting electrodes. This is achieved by having the electrolyte on a supporting medium such as paper or silica gel and is referred to as Zone Electrophoresis as opposed to Free Electrophoresis in a liquid (*see* Figure 7). A^+ and B^+ migrate to the negative electrode (cathode), C^- migrates to the positive electrode (anode) whilst neutral D remains at the origin.

An amino acid, which contains both acidic and basic groups can exist in different ionic forms depending on the pH of its solution and, therefore, the extent of migration and its direction depend on the pH of the buffer solution used.

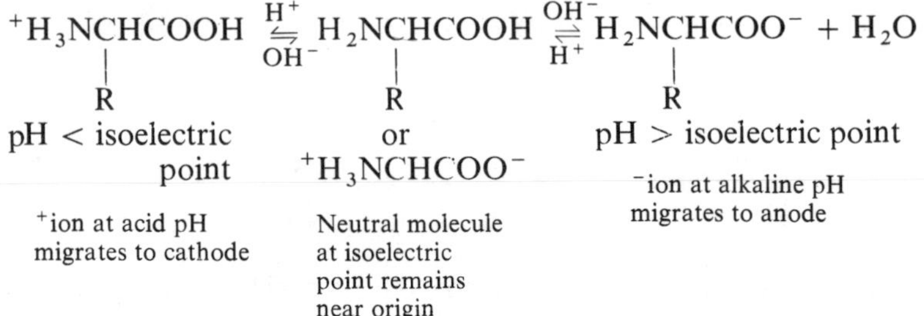

$^+H_3NCHCOOH \underset{OH^-}{\overset{H^+}{\rightleftharpoons}} H_2NCHCOOH \underset{H^+}{\overset{OH^-}{\rightleftharpoons}} H_2NCHCOO^- + H_2O$

R	R	R
pH < isoelectric point	or $^+H_3NCHCOO^-$	pH > isoelectric point
$^+$ion at acid pH migrates to cathode	Neutral molecule at isoelectric point remains near origin	$^-$ion at alkaline pH migrates to anode

If the side-group, R, contains an acidic or basic group this will
also affect the migration.

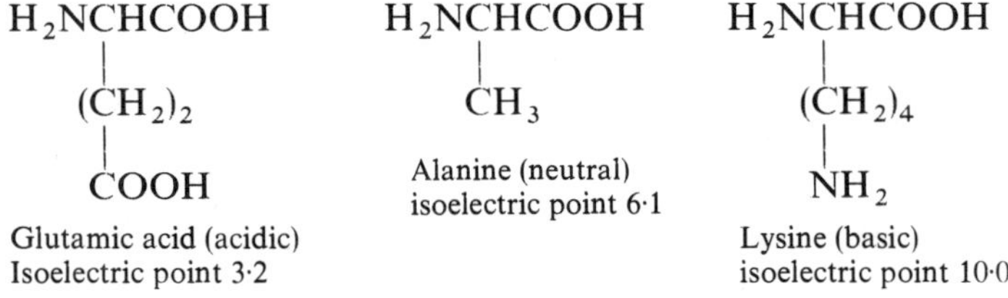

$H_2NCHCOOH$	$H_2NCHCOOH$	$H_2NCHCOOH$
$(CH_2)_2$	CH_3	$(CH_2)_4$
$COOH$	Alanine (neutral) isoelectric point 6·1	NH_2
Glutamic acid (acidic) Isoelectric point 3·2		Lysine (basic) isoelectric point 10·0

In a neutral solution, the acidic amino acids move towards the
anode, the basic amino acids move towards the cathode and the
neutral amino acids remain near the origin. If the solution is made
more acidic, the neutral amino acids will move towards the cathode

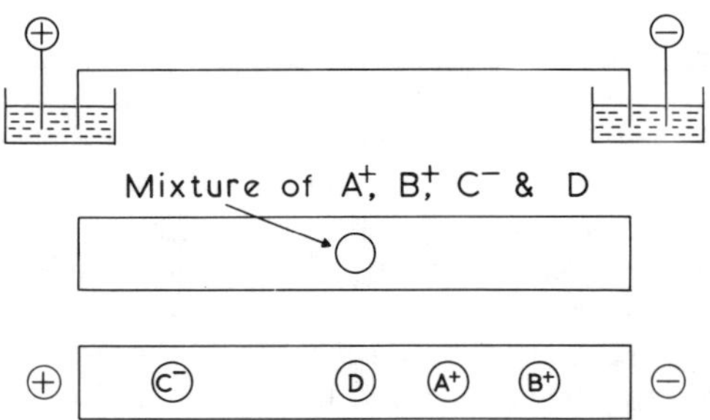

Figure 7. Principles of paper electrophoresis

and the ionization of the acidic amino acids will be suppressed so
that they will not move so far towards the anode and if the pH is
low enough, will actually migrate to the cathode. It is apparent,
therefore, that the extent and direction of travel depends on the

pH of the buffer solution used and the correct selection of buffer affords a predictable separating behaviour.

Although the theory of electrophoresis is simpler than that of chromatography, and foreign ions (salts) do not interfere with the former technique, the resolution is not as good as that obtained by chromatography because the amino acids tend to move in groups (acidic, basic and neutral) and the neutral amino acids rarely separate well. Furthermore, neutral compounds, for example glucose or an amino acid at its isoelectric point, move a slight distance from the origin towards the cathode due to the electro-osmotic effect (*see* Smith and Feinberg [1] p. 122).

It is possible to carry out two-way separations using electrophoresis in each direction but it is more advantageous to carry out chromatography in one direction and electrophoresis in the other. If the buffer used for the latter contains inorganic salts, the chromatography should be carried out first.

EXPERIMENTAL PROCEDURE

Prepare 0·2 per cent ninhydrin in acetone and approximately 0·01M-solutions of the amino acids in water alone or 10 per cent isopropanol–water (*see* page 30). The following buffer solutions are prepared by making up the quantities stated to 1 litre with water. N-acetic acid (60 g glacial acid), N-ammonia (71 ml 0·880 reagent), and pyridine–acetic acid (40 ml pyridine + 3·2 ml acetic acid). The pH of these solutions is 2·4, 11·6 and 6·1 respectively.

Use a 30 × 10 cm strip of Whatman No. 1 paper and draw a faint pencil line across the paper in the centre. Make pencil marks at 2 cm intervals along this line to serve as origins (*see* Figure 8a). Mark with the signs of the electrodes and label the origins. Wet the paper by dipping in a tray of the buffer solution to be used, blot between sheets of clean blotting paper and place the paper in position on the Shandon 'Unikit' electrophoresis assembly (Figure 8b). Alternatively, place the dry paper in position and allow the solution to soak up the paper (about 2 h). Add buffer solution to each compartment (about 50 ml) until it reaches the height of the groove on the electrode pillar, ensuring that the levels are equal to prevent siphoning.

Apply drops of the test solutions of alanine, aspartic acid, lysine and cysteic acid by means of melting-point tubing. Insert the assembly into the tank, replace the lid, connect to the power supply of approximately 300 V and allow to run for 2 h. At the end of the expt., turn of the current, lift out the assembly, remove the paper and blot the ends which were immersed in the buffer by touching on to clean filter paper. Hang the paper by stainless steel clips and dry

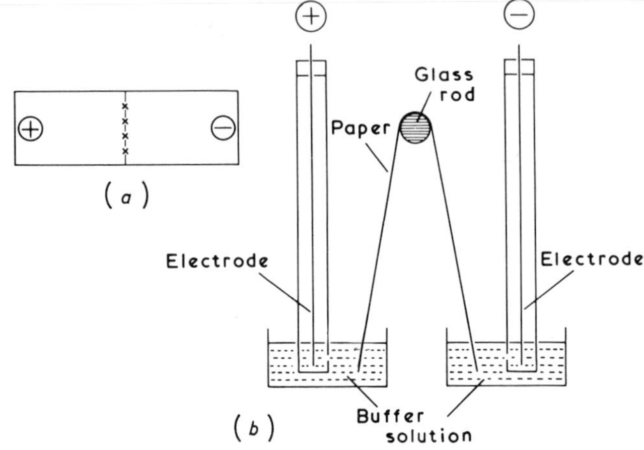

Figure 8.   Apparatus for paper electrophoresis

with a hair-dryer. Spray with ninhydrin and allow to stand at room temperature or heat for a few minutes at 100–110°C to locate the amino acids.

Use the three buffer solutions and correlate the results with theory. Finally, attempt to separate a mixture containing all four amino acids. It should be noted that the acid buffer solution suggested will not suppress the ionisation of the sulphonic acid group present in cysteic acid which has a very low isoelectric point.

REFERENCE
1 SMITH, I. and FEINBERG, J. G., *Paper and Thin-Layer Chromatography and Electrophoresis,* Shandon, London, 115 (1965)

14 Identification of N-terminal amino acids by the 'Dansyl' method

APPARATUS

Shandon 'Unikit' No. 2, including 10×10 cm glass plates and $10 \times 2 \cdot 5$ cm slides, ultra-violet lamp.

MATERIALS

1-(dimethylamino)-naphthalene-5-sulphonyl chloride, benzene, pyridine, glacial acetic acid, acetone, Kieselgel G, sodium bicarbonate, glycyl-DL-leucine and DL-leucyl-glycine (from Koch–Light Laboratories), glycine, DL-leucine.

THEORY

The usual technique for identifying N-terminal amino acids in peptides or proteins is to label the terminal amino acids, which contain a free amino group, by reaction with a compound of the type XY, where Y = reactive halogen. Provided the X—N bond is sufficiently stable, on hydrolysis the peptide will break down into its constituent amino acids, but the N-terminal acids will be modified and these can be characterised, usually by some form of chromatography. Dibasic amino acids, for example lysine, will react also by virtue of their side-chain amino groups, but these will be ω-substituted only, unless they are N-terminal when they will be α,ω-disubstituted. Reaction will occur also with the hydroxyl group of tyrosine side-chains.

The best known reagent for N-terminal amino acid determinations is 2:4-dinitrofluorobenzene, $2:4\text{-}(NO_2)_2C_6H_3F$, developed by Sanger

$$\text{XY} + \text{H}_2\overset{\displaystyle R}{\text{NCH}}\text{CONH}\overset{\displaystyle R^1}{\text{CH}}\text{CO---} \xrightarrow{\text{alkali}} \text{XHN}\overset{\displaystyle R}{\text{CH}}\text{CONH}\overset{\displaystyle R^1}{\text{CH}}\text{CO} \text{---} + \text{HY}$$

$$\text{XHN}\overset{\displaystyle R}{\text{CH}}\text{CONH}\overset{\displaystyle R^1}{\text{CH}}\text{CO} \xrightarrow[\text{(acid)}]{\text{hydrolysis}} \text{XHN}\overset{\displaystyle R}{\text{CH}}\text{COOH} + \text{H}_2\overset{\displaystyle R^1}{\text{NCH}}\text{COOH}$$

Labelled amino acid   Unchanged amino acids + etc.

during his studies on the structure of insulin, the N-terminal amino acids giving rise to yellow dinitrophenylated (DNP) derivatives of structure $2\!:\!4\text{-}(NO_2)_2C_6H_3NHCH(R)COOH$.

A more recent, but similar technique Gray and Hartley [1], uses

I
N(CH₃)₂ ... SO₂NH CHR COOH
DNS – Amino acid

II
N(CH₃)₂ ... SO₂Cl
DNS – chloride

'Dansyl' or DNS derivatives (I), formed by reaction with 1-(dimethyl-amino)-naphthalene-5-sulphonyl chloride (II), ('dansyl' chloride or DNS-chloride). These derivatives are very resistant to acid hydrolysis and can be detected on chromatograms by their fluorescence with a sensitivity 100 times greater than DNP-amino acids.

EXPERIMENTAL PROCEDURE

(*a*) N-*Terminal analysis of glycyl-*DL*-leucine and *DL*-*leucyl-glycine*—Prepare 0·1M sodium bicarbonate (8·4 g $NaHCO_3$ per litre) and DNS-chloride reagent (0·05 g DNS-Cl in 10 ml acetone).

Dissolve the peptide (1 mg approx.) in the sodium bicarbonate solution (1 ml) in a small test-tube and add the DNS-Cl in acetone (1 ml). Stopper and allow to stand for 3 h at room temperature in the dark (Note 1). Add acetone (5 ml) and filter off the precipitated sodium bicarbonate onto a small filter paper. Evaporate the solution to dryness in a dish in an evacuated desiccator over silica gel.

Add 5N-hydrochloric acid (2 ml) and transfer the solution of DNS-peptide to a drawn-out glass tube. Seal and place in an oven at 105–110°C overnight for hydrolysis to occur (*see* page ix). Allow to cool, break the neck of the tube and transfer the contents to a small dish or tube and evaporate to dryness in an evacuated desiccator over sodium hydroxide. Dissolve the residue in 50 per cent aq. acetone (1·0 ml) (*see* Note 2) for application to thin-layer plates (*see* page 33).

Use 10 × 2·5 cm thin-layer Kieselgel slides and a solvent (one way) consisting of benzene–pyridine–glacial acetic acid (40:10:1 by

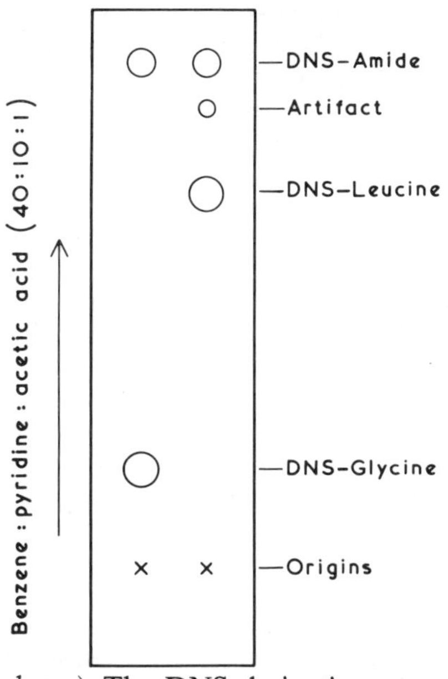

*Figure 9.   Separation of DNS-amino acids from glycyl-*DL-*leucine and* DL-*leucyl-glycine by one-way T.L.C.*

volume). The DNS derivatives appear as yellow fluorescent spots when the dried plate is examined under an ultra-violet lamp.

Figure 9 gives the relative positions of the spots obtainable. The positions of DNS-glycine and DNS-leucine from the respective dipeptides can be confirmed by the preparation of authentic DNS-derivatives of these amino acids (*see* Section (*c*)).

(*b*) N-*Terminal analysis of* α-*keratose*—Dissolve powdered α-keratose (*see* page 85) (50 mg) in 0·1M-sodium bicarbonate (2 ml), shaking if necessary and add DNS-chloride reagent in acetone (2 ml). Allow to stand for 3 h in the dark at room temperature. Acidify with 2N-hydrochloric acid and filter off the precipitated DNS-α-keratose on a small filter paper in a funnel. Wash with distilled water. Hydrolyse with 2 ml of 5N-hydrochloric acid, take down to dryness and dissolve in aq. acetone (0·1–0·5 ml) as described under (a).

For two-way chromatographic separation, use 10×10 cm thin-layer Kieselgel plates in the Shandon 'Unikit' No. 2 apparatus (*see* Note 1). As first solvent, use benzene–pyridine–glacial acetic acid (40:10:1 by volume) for 2 h. Allow to dry for 1 h at room temperature and then run for 1 h using as second solvent n-butanol saturated with 0·2N-sodium hydroxide. To prepare the latter solvent, shake n-butanol with an equal volume of 0·2N-sodium hydroxide (8 g/l) in a separating funnel and use the top organic layer. After development of the chromatogram, dry as before. The DNS derivatives appear as yellow fluorescent spots when the plate is examined under ultra-violet light. This should be carried out immediately, as the spots disappear if the plate is allowed to stand for a long period, particularly in the light.

Seven N-terminal amino acids have been found in α-keratose, aspartic and glutamic acids, serine, threonine, glycine, alanine and valine. In addition to these, the side-chains of some amino acid residues react with DNS-chloride and the derivatives formed, for example O-DNS-tyrosine (deep yellow), N(ε)-DNS-lysine and N(im)-DNS-histidine (orange) will be present in relatively high concentrations. Cysteic acid may be found also as an end-group.

By means of Figure 10 it should be possible to locate spots corresponding to these amino acids. Standard DNS-amino acids can be run along the side of the plate in the second direction only, to verify the positions of some of the spots.

(*c*) *Preparation of DNS-amino acids*—Dissolve the amino acid (approx. 1 mg) in 0·1M-sodium bicarbonate (2 ml). Add DNS-chloride reagent in acetone (2 ml) and allow to stand for 3 h at room temperature. Add acetone (10 ml) and filter off the precipitated sodium bicarbonate. Evaporate to dryness in an evacuated desiccator over silica gel and dissolve the residue in 50 per cent aq. acetone (2 ml) for chromatographic examination.

NOTES

1. DNS-amino acids are similar to DNP derivatives in that they are very susceptible to photodegradation. Wherever possible, if the material is allowed to stand, light should be excluded.

2. The quantities of 50 per cent aqueous acetone given to dissolve the DNS-amino acids should give fluorescent spots of reasonable intensity. If the spots on the plate are too weak, repeated applications of the solution to the plate should be made, allowing to dry

between applications. If the spots are too large and tail, the solution
to be applied should be diluted.

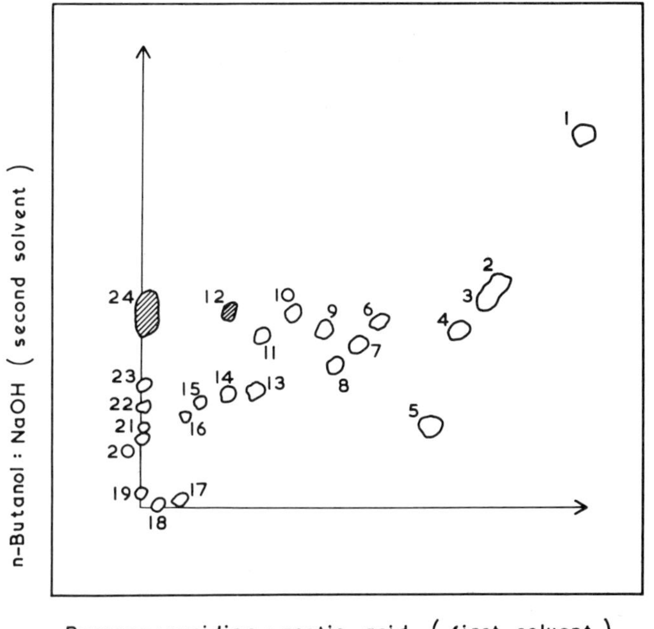

1. DNS-Amide	9. Di-DNS-Tyrosine	17. DNS-Glutamic acid
2. DNS-Isoleucine	10. Di-DNS-Lysine	18. DNS-Aspartic acid
3. DNS-Leucine	11. DNS-Tryptophan	19. DNS-Cysteic acid
4. DNS-Valine	12. DNS-Acid (Artifact, blue)	20. DNS-Arginine
5. DNS-Proline	13. DNS-Glycine	21. N(im)-DNS-Histidine (orange)
6. DNS-Phenylalanine	14 Di-DNS-Histidine	22. N(ε)-DNS-Lysine
7. DNS-Methionine	15. DNS-Threonine	23. O-DNS-Tyrosine (deep yellow)
8. DNS-Alanine	16. DNS-Serine	24. DNS-Acid (blue)

(Yellow fluorescence unless stated otherwise)

Figure 10. Separation of DNS-amino acids by two-way T.L.C.

REFERENCE

1 GRAY, W. R. and HARTLEY, B. S., *Biochem J.* **89**, 59 p. (1963)

15 The Spekker photo-electric absorptiometer

APPARATUS

Hilger Spekker photo-electric absorptiometer with matched cells and filters.

MATERIALS

Orange II, (also known as β-naphthol orange or Tropaeolin 000), industrial methylated spirit.

DISCUSSION

Colorimetric analysis is a method for determining the concentration of a substance in solution. The substance must be either inherently coloured or its solution must be capable of developing a colour by the addition of a suitable reagent. Use is made of the fact that the intensity of the colour of a solution increases with its concentration and thus by comparing its colour with standard solutions of known concentration, the amount of substance present in the original solution may be determined.

In photo-electric methods, the coloured solution is placed in front of a light source and the human eye is replaced by a photo-electric cell, the output of which affords a direct measure of the light intensity and hence of the absorption. It should be noted that all photo-electric methods measure light transmission and not, as in visual methods, the colour (scattered light) of the solution. In a one-cell instrument, a calibration curve is obtained by plotting the concentration of the solution placed in the light beam v. the galvanometer reading for a series of solutions of known concentration. The concentration of an unknown solution may then be determined from the galvanometer reading.

The Spekker photo-electric absorptiometer consists of a source of light situated midway between two photo-electric cells. Interposed between the light and each cell is a diaphragm, the latter being adjusted so that the current from the two cells balance. If an absorbing solution is now placed between one of the cells and the light, the corresponding diaphragm must be opened further to restore the balance of the cells and the extent of this opening, as measured on a calibrated drum, is related to the light absorption.

Although it is generally recognised that a two-cell instrument, such as the Spekker, employing a null-point method of balancing cells against each other is more accurate than a one-cell instrument, the latter, for example the 'Eel' absorptiometer, is more convenient to use if it is necessary to examine a large number of samples such as is obtained from an ion-exchange column and fraction collector.

This brief discussion is limited to the absorption of visible light, i.e. the examination of coloured solutions, but a more detailed account, including the use of ultra-violet and infra-red radiations, is given in the reference cited.

EXPERIMENTAL PROCEDURE

(*a*) *Method of use*—A simplified diagram of the Spekker photo-electric absorptiometer is shown in Figure 11 and its method of operation is as follows.

Switch on the lamp (L) by the switch at the back of the instrument. Select two glass cells (C) of suitable path length according to the optical densities of the solutions used. The cell size should be chosen so that the readings obtained over the range of concentrations used are spread over the complete scale. In general, a 2 cm cell will give twice the reading of a 1 cm cell. Set the sensitivity control (R) to minimum, ascertain that both heat absorbing filters (H) are in position and insert a pair of suitable coloured filters (F). The filter chosen should be the one which gives the steepest calibration curve, i.e. for a given concentration of the absorbing substance the drum reading should be a maximum. This may be determined by inserting each pair of filters into the instrument and ascertaining which filter gives the maximum reading for a given solution. Alternatively, a filter may be selected whose colour is as close as possible to the complementary colour of the solution, for example, a blue solution requires a red filter and vice versa. A table of complementary colours is given by Vogel [1] p. 757.

Fill one cell (C_1) with the coloured solution to be measured and the second cell (C_2) with the solvent or blank solution. It is preferable to mark and use the same cell each time for solution and blank

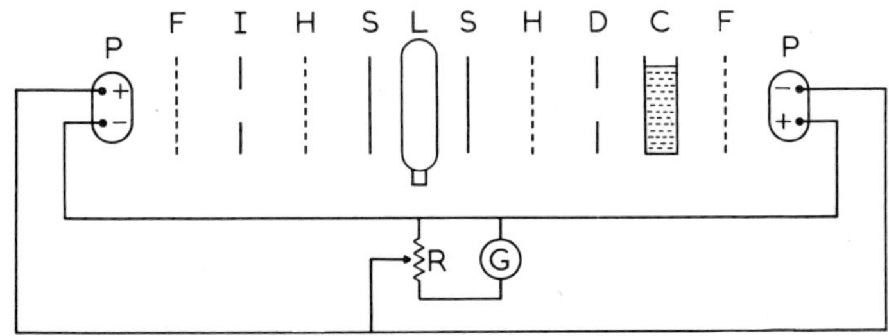

Figure 11.   Simplified diagram of the Spekker photo-electric absorptiometer

respectively in case the cells are not perfectly matched. Place the cells in position on the cell-mount as near the lamphouse as possible. Move the cell-mount so that the cell containing the coloured solution C_1' is in the light path. Set the drum (D) to read zero on the left-hand optical density scale. Close the iris diaphragm (I). Open the lamphouse shutter (S) and increase the sensitivity until there is about full-scale deflection of the galvanometer (G) spot. Bring the spot back to zero by opening the iris diaphragm (I), making the final adjustment with the fine rod adjustment. The light reaching the two photo cells (P) is now balanced.

The lamphouse shutter (S) is closed and the cell-mount moved so that the cell containing the blank solution C_2 is in the light path. Open the lamphouse shutter and bring the galvanometer spot back to zero by rotating the drum to a higher reading, i.e. decreasing the amount of light to compensate for the increased transmission of light by the blank over that of the solution. Close the lamphouse shutter and note the reading on the left-hand scale. It should be noted that since the intensity of light transmitted by a solution decreases exponentially as the concentration of the absorbing substance increases arithmetically, the adjustable diaphragm D is cam-shaped to give a linear scale on the drum.

(b) *Determination of dyes in solution*—Dissolve technical Orange II (5 g) in boiling water (50 ml). Filter through a fluted paper and add methylated spirit (150 ml) to the hot filtrate. Allow to cool, preferably allowing to stand overnight, filter under vacuum and wash with a little methylated spirit. The purified product, has the structure $p\text{-}NaO_3SC_6H_4N{=}NC_{10}H_6OH$ (β) after drying at 110°C in an oven for 3 h.

Weigh accurately 1·000 g purified Orange II and make up to 1 litre with distilled water in a volumetric flask. Measure accurately 100 ml of this solution and dil. to 1 litre in a volumetric flask. Take 10, 25, 50 and 75 ml portions of this solution and make each up to 100 ml with water in graduated flasks. These solutions contain (per ml) 10 µg, 50 µg and 75 µg of Orange II

Using the Spekker absorptiometer as described, with 1 cm glass cells, construct calibration curves (concentration along the x axis).

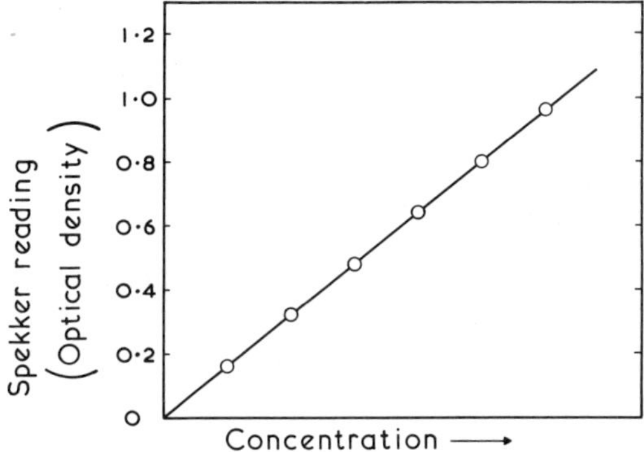

Figure 12.   Calibration curve for Spekker photoelectric absorptiometer

using a green (No. 4) filter and a yellow (No. 6) filter (*see* Figure 12). Comment on the graphs obtained and select the most suitable filter for subsequent use.

If wool is dyed with Orange II in acid solution (*see* page 70) placing a clock-glass over the beaker to avoid excessive evaporation (a suitable dye concentration is 2 per cent on the weight of material with a liquor ratio of 200:1), by removing samples of the dyebath and determining the concentration of dye with the absorptiometer, the degree of exhaustion at any given time may be found. The colour of the solution is independent of pH over the range 1–9. In the initial stages of dyeing, the solutions may require dilution in order to bring their concentration within the range of the calibration graph.

REFERENCE
1 VOGEL, A. I., *Textbook of Quantitative Inorganic Analysis.* Longmans, London, 738 (1961)

E

16 Determination of the cystine content of a protein

APPARATUS

Spekker photo-electric absorptiometer, hydrogen sulphide generator.

MATERIALS

Ortho-phosphoric acid (90 per cent w/w), mercuric chloride, sodium tungstate ($Na_2WO_4 2H_2O$), sodium metabisulphite ($Na_2S_2O_5$), lead acetate, copper sulphate, sodium acetate, acetic acid, cystine, ethyl alcohol (Industrial methylated spirits), wool samples.

THEORY

Sodium bisulphite reduces cystine as follows:

$$
\begin{array}{l}
\text{COOH} \\
| \\
\text{CH—CH}_2\text{—S—S—CH}_2\text{—CH} \quad\quad \text{COOH} \\
|\qquad\qquad\qquad\qquad\quad | \\
\text{NH}_2 \qquad\qquad\qquad\qquad\text{NH}_2
\end{array}
\quad + \text{ NaHSO}_3
$$

$$
\downarrow
$$

$$
\begin{array}{l}
\text{COOH} \qquad\qquad\qquad\qquad\qquad\qquad \text{COOH} \\
| \qquad\qquad\qquad\qquad\qquad\qquad\qquad\quad | \\
\text{CH—CH}_2\text{—SH} \quad + \text{ NaSO}_3\text{S—CH}_2\text{—CH} \\
| \qquad\qquad\qquad\qquad\qquad\qquad\qquad\quad | \\
\text{NH}_2 \qquad\qquad\qquad\qquad\qquad\qquad\qquad\text{NH}_2
\end{array}
$$

Cysteine   Sodium cysteine sulphonate

The cysteine gives a colour ('Tungsten blue') with the phosphotungstic acid reagent which is determined colorimetrically. Mercuric
chloride reacts with cysteine so that any colour produced in the
blank is not due to cysteine.

$$\overset{|}{\underset{|}{C}}H—CH_2—SH + HgCl_2 \rightarrow \overset{|}{\underset{|}{C}}H—CH_2—SHgCl + HCl$$

The sodium hydroxide solution neutralises the hydrochloric acid
used in the hydrolysis and the sodium acetate and acetic acid buffer
the solution at pH 5.

EXPERIMENTAL PROCEDURE

(a) *Preparation of phosphotungstic acid reagent*—Dissolve sodium
tungstate (25 g) in water (50 ml) and add phosphoric acid (5·0 ml)
slowly with cooling. Pass hydrogen sulphide through the solution at
a moderate rate for 20 min. (All operations involving hydrogen
sulphide should be carried out in a fume cupboard.) After the
hydrogen sulphide has been passing for 3–4 min, add more phosphoric acid (2·5 ml) without interrupting the gas flow. Filter the
solution, if necessary re-filtering the first few millilitres of the filtrate.
The filtrate should be clear and slightly green. Transfer to a 250 ml
separating funnel and add ethyl alcohol (75 ml) with shaking.
Remove the heavy bluish solution containing the phosphotungstic
acid, which separates at the bottom, rather quickly into a tared
250 ml round bottomed flask with a ground neck. Any solid
aggregates floating at the interface of the two layers should not be
allowed to pass through the stopcock. Add water until the weight
of the solution is 75 g and boil over a small flame until there is no
reaction for evolved hydrogen sulphide with lead acetate paper.
Reduce the flame, add phosphoric acid (5·0 ml), fit a water condenser to the flask and gently reflux the solution for 1 h. Remove
the condenser, add a few drops of bromine water to remove any
blue colour and boil rapidly to remove excess of bromine. After
cooling, transfer contents and washings to a 250 ml graduated flask,
add $CuSO_4 \cdot 5H_2O$ (0·25 mg) or 0·1 per cent $CuSO_4 \cdot 5H_2O$ solution
(0·25 ml) and finally make up to 250 ml with water and mix
thoroughly.
(b) *Calibration of Spekker photo-electric absorptiometer*—The use of
this instrument is described on page 48.

The following solutions are prepared by making up the quantity stated to 250 ml with water:

M-Sodium hydroxide	10·0 g of pellets
2M-Sodium acetate	41·0 g of anhydrous salt or 68·0 g of hydrated ($3H_2O$) material
2M-Acetic acid	30·0 g (29·0 ml) of glacial acid
M-Sodium metabisulphite	48·0 g of $Na_2S_2O_5$ and 1·0 g of sodium hydroxide
0·10M-mercuric chloride	9·3 g of $HgCl_2$

5N-Hydrochloric acid is prepared by diluting 430 ml of conc. acid to 1 litre with water.

Standard cystine solutions are made by dissolving accurately known weights of approximately 0·2, 0·3, 0·4, 0·5 and 0·6 g of pure dry cystine in 5N-HCl (80 ml) in graduated flasks. Make up to 1 litre with water and shake well to ensure uniform mixing. If a semi-micro balance is available, weights of cystine ranging from 20 to 60 mg may be dissolved in 8·0 ml of 5N-NCl and made up to 100 ml.

Into each of two 100 ml graduated flasks (A and B) place, in the order given, M-sodium hydroxide (2·0 ml), 2M-sodium acetate (10·0 ml), 2M-acetic acid (3·0 ml), standard cystine solution (5·0 ml) and freshly prepared M-sodium metabisulphite (3·0 ml). To solution B only, i.e. the blank, add 0·10M-mercuric chloride (3·0 ml). Add water (25·0 ml) to each solution and allow to stand for 15 min. Then add the phosphotungstic acid reagent (3·0 ml) to each solution. Allow to stand for 20 min, dilute to 100 ml and measure solution A colorimetrically using the Spekker photo-electric absorptiometer with red (No. 8) filters and 1 cm glass cells against solution B as the blank.

Construct a calibration chart of Spekker reading against concentration of cystine (mg/ml).

(c) *Cystine analysis of wool*—Approx. 0·25 g of wool at atmospheric regain is dried by spreading out in an oven for 1 h at 110°C or in an evacuated desiccator over phosphorus pentoxide for 2 days. The dry weight, which should be between 0·18 and 0·22 g, is accurately determined. The wool is hydrolysed in a sealed tube with 5N-HCl (4·0 ml) overnight in an oven at 110°C (*see* page ix). After allowing to cool, the tube is broken open and the contents filtered through a small paper into a 50 ml graduated flask. The tube and paper are washed several times with water and the hydrolysate and washings are finally made up to 50 ml. The procedure for calibration is now carried out using this hydrolysate solution instead of the cystine solutions of known composition. From the calibration graph, the amount of cystine present in the hydrolysate is determined and this is expressed as a percentage of the original dry weight of wool.

Cystine determinations may be carried out on the following:

(i) Australian wools (after removal of tip-ends) of different qualities, for example 80s, 64s and 40s. The purification procedure for greasy wools is given on page 86.
(ii) Wool fabrics before and after an oxidative bleaching or anti-felting treatment.
(iii) The root and tip-ends of badly weathered wool fibres.

NOTE

To carry out the determination on a semi-micro scale, the same procedures and reagents are used, but quantities are modified as follows:

Calibration of absorptiometer—Dissolve known weights (ranging from 20 to 60 mg) of pure dry cystine in 5N-hydrochloric acid (40 ml) and make up to 1 litre with water. Into each of two 25 ml graduated flasks (A and B) place, in the order given, M-sodium hydroxide (2·0 ml), 2M-sodium acetate (2·0 ml), 2M-acetic acid (0·6 ml), cystine solution (10·0 ml) and M-sodium metabisulphite (1·0 ml). To solution B only (the blank) add 0·1M-mercuric chloride (1·0 ml).

Add distilled water (5·0 ml) to each solution and allow to stand for 15 min. Then add phosphotungstic acid reagent (1·0 ml) to each. Allow to stand for 20 min, dil. to 25 ml and measure colorimetrically with the Spekker absorptiometer to obtain the calibration graph.

Determination of cystine in wool—Hydrolyse an accurately known weight (approx. 10 mg) of the dry wool with 5N-hydrochloric acid (1·0 ml) filter and make up to 25 ml with distilled water in a graduated flask. Using the new calibration curve, determine the cystine content as described for the macro method.

REFERENCES
1 FOLIN, O. and MARENZI, A. D., *J. Biol. Chem.,* **83**, 109 (1929)
2 SHINOHARA, K., *J. Biol. Chem.,* **109**, 665 (1935)
3 ALEXANDER, P., HUDSON, R. F. and FOX, M., *Biochem J.,* **46**, 27 (1950)

17 Determination of the tyrosine content of a protein

APPARATUS

Spekker photo-electric absorptiometer. No. 3 porosity sintered-glass funnels.

MATERIALS

Mercuric sulphate, mercuric chloride, anhydrous sodium sulphate, sodium nitrite, tyrosine, samples of wool and silk.

THEORY

Millon's reagent (a solution of mercury in nitric acid) for phenols was employed by Hoffmann in 1853 for testing for tyrosine. The test depends on the presence of a little nitrous acid in the solution and it is generally accepted that the red colour is due to chelation between the nitrosophenol and the mercury.

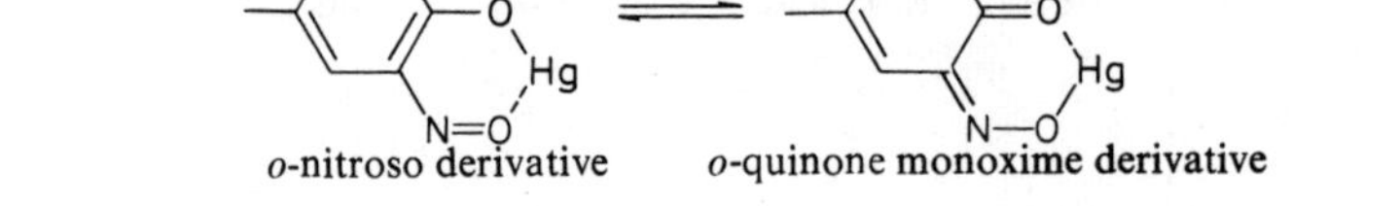

$$\text{—}\langle\text{C}_6\text{H}_4\rangle\text{OH} + HNO_2 + \tfrac{1}{2}Hg_2Cl_2 \longrightarrow \quad + H_2O + HCl$$

The *o*-nitroso derivative probably exists in equilibrium with its tautomeric form, the *o*-quinone monoxime derivative.

o-nitroso derivative   o-quinone monoxime derivative

EXPERIMENTAL PROCEDURE

(*a*) *Preparation of reagents*—M-sodium nitrite is prepared by making up $NaNO_2$ (17·2 g) to 250 ml with water, and 0·5M-sulphuric acid by adding 27 ml of the conc. acid to water and after cooling, making up to 1 litre.

Reagent A is made by dissolving mercuric sulphate (75 g), mercuric chloride (55 g) and anhyd. sodium sulphate (70 g) in water (850 ml) and adding carefully with agitation conc. sulphuric acid (125 g or 68 ml). When the solution has attained room temperature, it is finally diluted to 1 litre with water.

Reagent B is prepared by diluting a volume of reagent A with an equal volume of 0·5 M-sulphuric acid.

(*b*) *Calibration of Spekker photo-electric absorptiometer*—5-N hydrochloric acid is prepared by diluting 430 ml of the conc. acid to 1 litre with water. Standard tyrosine solutions are made by dissolving accurately known weights ranging from 0·1 g to 0·6 g (0·1 g increments) of pure dry tyrosine in 5-N HCl (80 ml) and making up to 1 litre in graduated flasks. The flasks are shaken well to ensure uniform mixing. If a semi-micro balance is available, weights ranging from 10 to 60 mg may be dissolved in 5N-HCl (8·0 ml) and made up to 100 ml.

Take exactly 10·0 ml (pipette) of a standard tyrosine solution, 4·0 ml of 0·5M-sulphuric acid and 10·0 ml of reagent A and heat to 60–65°C for 30 min in a water bath. Cool for 1 h at a few degrees below room temperature and filter through a No. 3 porosity sintered-glass funnel into a 50 ml graduated flask (This is necessary to remove any cloudiness from the test samples due to residual pigment and tryptophan, and is not strictly necessary for the standardising solutions, but it is advisable to carry out the same procedure on the calibration samples.) Wash the flask and filter with reagent B and finally make the solution up to the mark with the same reagent.

Take 8·0 ml of 5N-HCl and make up to 100 ml with water in a graduated flask. Using 10·0 ml of this solution in place of the standard tyrosine solution, prepare a blank using reagents A and B exactly as described previously.

To each solution (blank and standard tyrosine solution) add 1·0 ml of M-sodium nitrite solution (freshly prepared) so that it floats on the surface. Shake the flasks together for exactly 3 min and then measure the colour using the Spekker photo-electric absorptiometer with No. 2 blue filters and 1 cm glass cells (for details *see* page 48).

Plot the graph of Spekker reading (optical density) versus tyrosine concentration (mg/ml).

If the tyrosine content of wool is being determined, it is found that the presence of cystine gives a result about 12 per cent too high. To correct for this, the standard tyrosine solutions should contain cystine equal in weight to twice that of the tyrosine present.

(*c*) *Tyrosine analysis of wool or silk*—Hydrolyse an accurately known weight (approx. 0·2 g dry) of silk or wool and make the hydrolysate up to 50 ml as described on page 54. Carry out the procedure as for the calibration using this hydrolysate solution instead of the standard tyrosine solutions. From the results, determine the amount of tyrosine present in the hydrolysate and express it as a percentage of the original dry weight of the wool or silk.

Tyrosine determinations may be carried out on the following:

(i) 64s Merino wool. Commercially purified material is satisfactory.

(ii) *Bombyx mori* silk. Again commercial silk may be used. If the silk is in the raw state, the gum (sericin) may be removed from the filaments (fibroin) by boiling for 30 min in two changes of 1 per cent soap solution followed by washing in water and drying.

(iii) Various semi-cultivated and wild silks. The amino acid composition of these may be quite different from *Bombyx mori*. They may be degummed in soap solution but their hydrolysis may present difficulties, for example Tussah. If after 8 h hydrolysis, appreciable fibrous material remains, the tube should not be unsealed, but hydrolysis continued for a further 8 h. After this period, it is possible some insoluble pigment may be produced. The hydrolysate should be filtered through a tared sintered glass crucible and the original weight of silk corrected for this residue.

(iv) Silk before and after degumming. If the weight loss on degumming is determined, the tyrosine content of both the fibroin and the sericin by difference may be found.

REFERENCES
1 Lugg, J. W. H., *Biochem J.*, **31**, 1422 (1937)
2 Nilssen, B., *Symp. Fibrous Proteins*, (publ. Soc. Dy. Col., Bradford), 142 (1946)

PART 2 Experiments on Proteins

18 The milling of wool fabrics

APPARATUS

No special apparatus is essential but it is preferable for the milling solutions to be contained in polythene bowls or stainless steel beakers or dye pots of about 500 ml capacity.

MATERIALS

Cadmium sulphide, acetone-soluble cellulose acetate (for example from Hopkin and Williams), acetone, soap flakes, sodium dihydrogen phosphate ($NaH_2PO_4 \cdot 2H_2O$), disodium hydrogen phosphate ($Na_2HPO_4 \cdot 12H_2O$) scoured knitted and woven wool fabrics, including tight and loose constructions.

THEORY

All fabrics, natural and synthetic, will undergo a loss in dimensions on immersion in water if they have been dried previously under tension. This is due to the release of strains and the contraction is known as relaxation shrinkage.

Fabrics of wool and related animal fibres, however, even if previously relaxed, shrink further when subjected to alternate compression and relaxation in aq. media. This contraction, which is known as milling or felting shrinkage, is due to the migration of the fibres and results from their unique surface structure and elasticity. A loosely constructed knitted fabric may lose up to 50 per cent of its area on milling. The phenomenon of milling or felting is complex and is discussed in greater detail in the references cited.

EXPERIMENTAL PROCEDURE

(*a*) *Preparation of patterns*—In a wide-topped reagent bottle, place cadmium sulphide (1·0 g), cellulose acetate (12·5 g) and acetone (100 ml). Preferably allow to stand overnight and stir before use.

The wool fabric used should be free from relaxation shrinkage and if necessary should be immersed in cold water containing a few drops of 'Teepol' for at least 1 h, followed by air drying without tension. Cut the fabric into rectangular patterns approximately 13 cm × 16 cm and by means of a glass rod, coat the edges with the acetate marking fluid to prevent fraying. Make datum marks on the

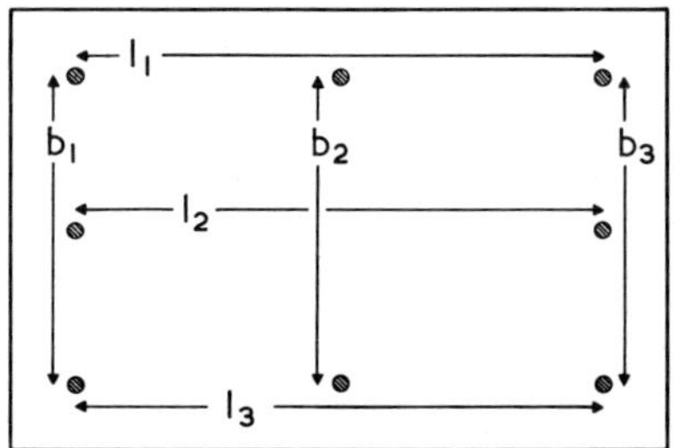

Figure 13.   Marking of pattern of fabric for shrinkage testing

fabric as shown in Figure 13. For identification purposes it is also convenient to number the patterns.

(*b*) *Preparation of Milling Solutions*—One per cent soap solution is prepared by dissolving soap flakes (10 g) in hot water and diluting to 1 litre with cold water. If the tap water is excessively hard, use distilled water or add sodium carbonate (approx. 0·2 g).

To prepare 0·1N hydrochloric acid solution, dilute conc. acid (9 ml) to 1 litre with water.

A buffer solution of pH 6 is made by dissolving $NaH_2PO_4 \cdot 2H_2O$ (15·6 g) and $Na_2HPO_4 \cdot 12H_2O$ (14·3 g) in water and making up to 1 litre to give a solution which is 0·1M with respect to the monosodium salt and 0·04M with respect to the disodium salt.

(*c*) *Determination of Milling Shrinkage*—Warm the soap solution to 30°C and maintain at this temperature throughout the milling. Immerse one knitted and one woven pattern in the soap solution for 5 min to thoroughly wet out. Remove from the soap solution, lay flat on a tile and measure, preferably with a steel ruler, the distance between the datum marks to obtain a value for the initial area.

If the length between the three pairs of marks is l_1, l_2 and l_3 cm and the breadth b_1, b_2 and b_3 cm,

$$\text{Initial area} = \left(\frac{l_1 + l_2 + l_3}{3}\right)\left(\frac{b_1 + b_2 + b_3}{3}\right) \text{ cm}^2$$

Hand-mill (*see* Note 1) the two samples together in the soap solution for 4 min and again measure.

$$\text{Percentage area shrinkage} = \frac{(\text{Initial area} - \text{final area})}{\text{initial area}} \times 100$$

Repeat the hand-milling for further periods of 4 min and measure again until the knitted pattern shrinks 40–50 per cent or a terminal value is obtained. Tabulate the results and plot a graph of time of milling versus percentage area shrinkage for each pattern.

Repeat using 0·10N-hydrochloric acid and pH 6 buffer solution also at 30°C as the milling media. Plot the six sets of results on the same axes for comparison and draw conclusions.

NOTES

1. In hand-milling, the patterns are squeezed *very vigorously* as they are passed from hand to hand. The actual milling takes place outside the milling medium but the patterns are wetted out about every 30 s. If the milling is carried out effectively, a loosely knitted fabric should shrink at least 30 per cent in 10 min.

2. The scope of this investigation can be extended by milling together knitted or woven fabrics of different constructions. Also by using milling media of differing temperatures, for example soap solution at 20°C and 50°C on patterns of the same fabric, the general effect of temperature can be determined. It must be emphasised that all these comparative results are valid only if all the milling is performed by the same operator.

REFERENCES

1 FRENEY, M. R., *Fibrous Proteins Symposium*. Soc. Dy. Col. Bradford, 178 (1946)
2 ALEXANDER, P., HUDSON, R. F. and EARLAND, C., *Wool—Its Chemistry and Physics*. Chapman & Hall, London, 38 (1963)
3 GORDON COOK, J., *Handbook of Textile Fibres*, vol. I. Merrow, Watford, England, 114 (1968)

19 Supercontraction of wool

APPARATUS

No special apparatus is required.

MATERIALS

Sodium metabisulphite ($Na_2S_2O_5$), lithium bromide, scoured knitted wool fabric, cadmium sulphide–cellulose acetate marking fluid (*see* page 62).

THEORY

The macromolecular structure of keratin is stabilised by disulphide cross-linkages and hydrogen bonds. Upon treatment of wool fibres with sodium bisulphite which ruptures disulphide bonds or conc. solutions of lithium bromide which break hydrogen bonds, these stabilising influences are removed and the fibres contract. Whereas the contraction brought about by disulphide bond breaking reagents is irreversible, with hydrogen bond breaking agents the fibre assumes its original length on washing in water. If treatment with lithium bromide solutions, however, is prolonged a disulphide-thiol interchange reaction occurs and the fibre becomes stabilised in its contracted state. The phenomenon of supercontraction is very complex in that the most effective reagent for breaking disulphide bonds in wool is chlorine which, in point of fact, does not produce supercontraction.

The term 'supercontraction' is retained on historical grounds in that the original work was performed on stretched fibres which contracted beyond their initial length. In these experiments 'super-contraction' is synonymous with simple contraction.

EXPERIMENTAL PROCEDURE

Cut the wool fabric into ten patterns approx. 5 cm × 5 cm. Edge with marking fluid and make a datum dot in each corner (*see* page 62). (*a*) *Supercontraction in Sodium Bisulphite Solution*—Make up sodium metabisulphite (20 g) to 100 ml with water and shake well to dissolve. The solution contains sodium bisulphite ($Na_2S_2O_5 + H_2O \rightleftharpoons 2NaHSO_3$). Take two patterns of wool fabric, wet out for 5 min in water containing a few drops of 'Teepol' and squeeze as dry as possible. Determine the mean dimensions. Take two boiling tubes, half fill each with sodium bisulphite solution, and immerse in a 600 ml beaker containing boiling water. When the bisulphite solution has reached thermal equilibrium, add the wool samples to the tubes and continue heating for 30 min.

Remove the duplicate samples and measure without rinsing. Measure again after standing in cold water for 30 min with frequent changing of the water. Record the contraction as a percentage of the original area both before and after washing. As an approximation, the linear contraction may be taken as half that of the area contraction.

(*b*) *Supercontraction in Lithium Bromide Solutions*—Make up four solutions of lithium bromide containing (a) 25 g LiBr + 25 ml water (b) 20 g LiBr + 30 ml water (c) 10 g LiBr + 40 ml water and (d) 5 g LiBr + 45 ml water. Place 20–25 ml of each solution in a boiling tube and heat the four tubes in boiling water contained in a 600 ml beaker until their contents are at approximately 100°C. Immerse one wetted out and measured sample in each solution for 1 min only, remove and without washing again measure. Allow the samples to stand in beakers of cold water for 30 min, changing the water occasionally and then measure again. Repeat the experiments on fresh patterns of fabric in the same lithium bromide solutions but treat for 30 min, measuring and washing after treatment as before.

Tabulate the percentage contraction in area of the eight patterns both before and after washing and correlate the results with the accepted views on supercontraction.

REFERENCES
1 ALEXANDER, P., HUDSON, R. F. and EARLAND, C., *Wool—Its Chemistry and Physics*. Chapman & Hall, London, 73 (1963)
2 EARLAND, C., *Review of Textile Progress*. Butterworths, London, 86 (1964)

20   The acid combining value of untreated and acetylated wool

APPARATUS

Mechanical shaker (preferable but not essential).

MATERIALS

0·05N-sodium hydroxide (exact normality required), 0·05N-hydrochloric acid containing 1·0 ml 'Teepol' detergent per litre (approximate normality only is necessary) (*see* Note), phenolphthalein indicator, acetic anhydride, glacial acetic acid, *N*,*N*-dimethylaniline, scoured wool (fibres, yarn or if in fabric form, of open structure, preferably knitted).

THEORY (*See* Alexander *et al.* [1])

Wool combines with acids due to salt formation with the basic groups present

$$\text{Wool–NH}_2 + \text{H}^+\text{X}^- \rightarrow \text{Wool–NH}_3^+\text{X}^-$$

For simple acids, the uptake is about 82 m.equiv./100 g of wool.

If the basic groups are blocked, for example by acetylation with acetic anhydride

$$\text{Wool–NH}_2 + \begin{array}{c} \text{CH}_3\text{CO} \\ \diagdown \\ \quad\quad\text{O} \\ \diagup \\ \text{CH}_3\text{CO} \end{array} \rightarrow \text{Wool–NHCOCH}_3 + \text{CH}_3\text{COOH}$$

then the acid absorption will be reduced to that corresponding to the degree of acetylation produced.

In point of fact, wool also contains carboxyl groups approximately equal in number to the basic groups, so that reaction with acid is usually regarded as a back titration of the Zwitterion structure.

$$\text{Wool–COO}^- \;{}^+\text{H}_3\text{N–Wool} + \text{H}^+\text{X}^- \rightarrow \text{Wool–COOH}$$
$$+ \text{Wool–NH}_3^+\text{Cl}^-$$

It must, however, be emphasised that the total amount of acid combined is determined by the number of basic groups present and the number of carboxyl groups is immaterial.

EXPERIMENTAL PROCEDURE

(a) *Acetylation of Wool*—The following reagents are mixed and cooled to room temperature: acetic anhydride (6·5 ml), glacial acetic acid (18·5 ml), conc. sulphuric acid (0·4 ml), *N*,*N*-dimethyl-aniline (0·4 ml). Cut the wool fabric (approx. 5·7 g equilibrated under laboratory conditions) into approximately 5 cm squares and dry for 1 h at 110°C. Place the dried wool in a 100 ml flask, add the mixed reagents and close the flask with a glass stopper. With occasional agitation, heat to 57–58°C over a period of 30 min on a water bath and allow the reaction to proceed at this temperature for a further 60 min. The temperature given, which is that of the wool, is critical. (CAUTION—should the acetylation mixture come into contact with the skin, wash thoroughly with soap and water.) Remove the wool, wash very thoroughly with cold tap water and finally distilled water and dry at room temperature.

(b) *Acid Absorption*—It is essential that the wool used for this determination, both untreated and acetylated, has been washed in distilled water until all traces of free alkali or acid have been removed. This may be determined by allowing a sample of the wool to stand for 10 min in distilled water with frequent agitation and noting if there is any change in the pH of the water (Universal Indicator).

Dry approximately 2 g samples of scoured wool for 24 h in a vacuum desiccator over phosphorus pentoxide and accurately weigh. If fibres or yarn are used, cut into 1 cm lengths, and if fabric, cut into 1 cm squares, and immerse in 100 ml (accurately measured) of 0·05N hydrochloric acid containing detergent in a dry 250 ml stoppered flask. (Carry out in duplicate.) Agitate frequently or

F

preferably place in a mechanical shaker for 1 h. Withdraw 25 ml portions of the liquid, avoiding fibres, and titrate against standard 0·05N sodium hydroxide solution using a few drops of phenolphthalein solution as indicator. Repeat the procedure on a known dry weight of acetylated wool. Finally, titrate 25 ml portions of the original acid against the standard alkali.

CALCULATION AND RESULTS

If the weight of wool taken is w g, and the difference between the titrations (on 25 ml) is x ml of sodium hydroxide solution (normality = yN).

Then, w g of wool absorb $4xy$ ml of N-acid.

w g of wool absorb $4xy$ m.equiv. of acid.

100 g of wool absorb $\dfrac{400xy}{w}$ m.equiv.

Acid combining value $= \dfrac{400xy}{w}$ m.equiv./100 g.

NOTE

0·05N Sodium hydroxide is prepared by making up approximately 2·2 g of the solid to 1 litre with water and standardising against a suitable acid (*see* page 73 and Vogel [2]). 0·05N Hydrochloric acid is made by diluting 4·0 ml of the conc. acid to 1 litre and the exact normality is not required.

REFERENCES
1 ALEXANDER, P., HUDSON, R. F. and EARLAND, C., *Wool—Its Chemistry and Physics*. Chapman & Hall, London, 18 and 315 (1963)
2 VOGEL, A. I., *Quantitative Inorganic Analysis*. Longmans, London, 243 (1961)

21 Rendering wool resistant to acid dyeing

APPARATUS

No special apparatus is required.

MATERIALS

Glyoxal-bis-sodium bisulphite, glyoxal, industrial methylated spirit, glacial acetic acid, acetic anhydride, *N*,*N*-dimethylaniline, sodium sulphate (anhydrous or hydrated), 'Teepol' detergent, an acid dyestuff, for example Carmoisine LS 2 g (approx.) patterns of scoured wool fabric.

THEORY

Nearly all acid dyes are sodium salts of sulphonic acid derivatives. When these dyes are applied in acid solution, their uptake by protein fibres is analogous to the absorption of simple acids (*see* page 66)

$$-COO^- {}^+H_3N- + H^+D^- \rightarrow -COOH + -NH_3^+D^-$$

where D^- represents the coloured anion.

For simple dyes, the uptake corresponds to the acid-combining capacity of the protein, for example 0·8–0·9 m.equiv./g in the case of wool, which is good evidence in favour of the mechanism suggested. It must, however, be pointed out that complex anions, including some dyestuffs, often combine far in excess of the figures quoted due to their own specific affinity for the fibre.

It is clear that if the amino groups provide the sites at which the dye anions are attached, the blocking of these groups by chemical modification will reduce the affinity of the fibre for the dyestuff.

In this experiment, amino groups are blocked by acetylation or by reaction with glyoxal-mono-sodium bisulphite

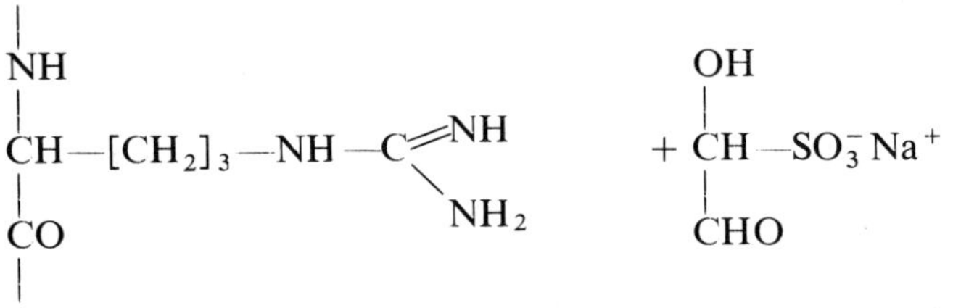

$$\underset{\text{arginine side-chain}}{\left| \begin{array}{l} NH \\ | \\ CH-[CH_2]_3-NH-C{\displaystyle \mathop{\big\langle}^{NH}_{NH_2}} \\ | \\ CO \\ | \end{array} \right.} \qquad + \underset{\text{glyoxal mono-sodium bisulphite}}{\left| \begin{array}{l} OH \\ | \\ CH-SO_3^-Na^+ \\ | \\ CHO \end{array} \right.}$$

$$\rightarrow \begin{array}{l} | \\ NH \\ | \\ CH-[CH_2]_3-NH-C{\displaystyle \mathop{\big\langle}^{N-CH-SO_3^-Na^+}_{N=CH}} \\ | \\ CO \\ | \end{array} + 2H_2O$$

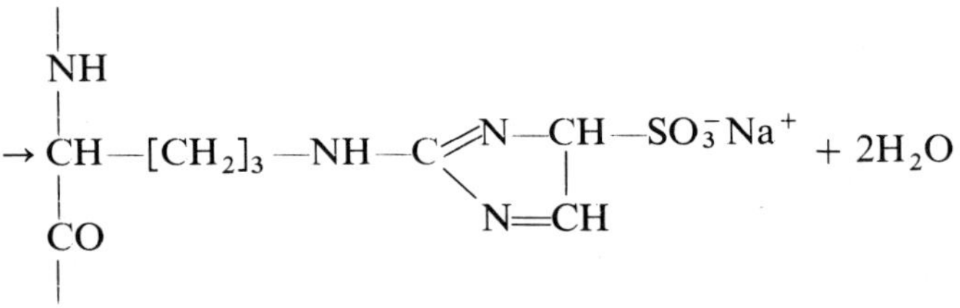

The arginine side-chains are replaced by groups, which not only have no affinity for dye anions, but will actually repel them.

It should be noted that the reaction is carried out using a mixture of glyoxal-bis-sodium bisulphite and glyoxal which is equivalent to the mono adduct.

EXPERIMENTAL PROCEDURE

(*a*) *Acetylation of Wool*—A 2 g pattern of scoured wool fabric is acetylated as described on page 67. It is washed off thoroughly in tap water until the washings are neutral, but need not be washed in distilled water nor dried.

(*b*) *Glyoxal Resist Treatment*—A 2 g pattern of scoured wool fabric is wetted out in water containing a few drops of 'Teepol' for 5 min. It is then rinsed and immersed for 1 h at 85°C in a solution consisting of water (180 ml), glyoxal-bis-sodium bisulphite (5 g), 30 per cent glyoxal solution (3·0 ml) industrial methylated spirit (10 ml), and glacial acetic acid (10 ml). The fabric is then rinsed well in water but not dried.

(*c*) *Acid Dyeing of the Patterns*—Take the two treated patterns and a

wetted-out untreated pattern (6 g of wool) and place in a 600 ml
beaker containing water (400 ml), anhydrous sodium sulphate
(0·6 g) or hydrated sodium sulphate (1·2 g) and glacial acetic acid
(0·2 ml) (*see* Note 1). Heat the bath slowly to 60°C, occasionally
stirring the patterns. Lift the patterns clear of the liquid with a
glass rod, add a 0·20 per cent solution of Carmoisine LS (6·0 ml)
(*see* Note 2), agitate thoroughly and return the patterns. Raise to
the boil and boil for 45 min. Rinse, dry and mount the samples for
comparison.

NOTES

1 This procedure is to ensure that all patterns are at the same pH
before dyeing commences, otherwise the 'initial strike' of dyestuff
may differ from pattern to pattern.
2 If the volume (in millilitres) of this solution added to the bath
is equal to the total weight (in grammes) of the wool, then the weight
of dye corresponds to 0·2 per cent on the weight of substrate. Other
suitable acid dyes may be used.

REFERENCES
1 ELLIOTT, J. H. and SPEAKMAN, J. B., *J. Soc. Dyers, Cols.* **59**, 185 (1943)
2 ALEXANDER, P., HUDSON, R. F. and EARLAND, C., *Wool—Its Chemistry and
Physics*. Chapman & Hall, London. 315 (1963)

22 The alkali solubility of wool

APPARATUS

No. 1 porosity sintered glass filtering crucibles, thermostat bath (preferable but not essential).

MATERIALS

0.100 ± 0.005N sodium hydroxide. This may be obtained in ampoules requiring dilution or pellet sodium hydroxide and a suitable acid for standardisation, for example potassium hydrogen phthalate, together with phenolphthalein indicator are required. Acetic acid, sodium hypochlorite solution, 100 vol. hydrogen peroxide solution (CAUTION—handle with care), scoured wool which may be loose or in the form of sliver, yarn or fabric.

DISCUSSION (*see* Alexander *et al.* [1])

Wool is insoluble in all non-destructive solvents. It is, however, soluble in caustic alkali due to hydrolytic breakdown of its macro-molecular structure. If treatment with sodium hydroxide is carried out under closely controlled conditions as described in this expt, depending somewhat on its quality, physical form, etc., approximately 10 per cent of the wool will dissolve. This quantity which dissolves under arbitrary conditions, is known as the 'alkali solubility'. If prior to the test, the wool has received chemical modification (damage) by treatment with acids, oxidizing or reducing agents, radiation, etc., its solubility in alkali increases. The alkali solubility test provides a very useful measure of the extent of chemical damage which may be due to either peptide bond breakdown or rupture of disulphide bonds. Conversely, the introduction of new cross-linkages into the keratin structure, for example with formaldehyde (*see* page 79) reduces its alkali solubility. It must be appreciated that alkali itself may replace the disulphide cross-linkages by the more stable lanthionine residues (*see* page 81) so that alkali solubility cannot be used to assess damage due to peptide-bond hydrolysis occurring in alkaline treatments. Finally, it should be emphasised that wherever possible, the test should be performed in conjunction with an untreated sample of the same material.

EXPERIMENTAL PROCEDURE

(*a*) *Preparation of* 0·100N-*Sodium Hydroxide*—Weigh out approximately 4·6 g of sodium hydroxide pellets on a watch glass and make up to 1 litre in a volumetric flask. Take about 2 g of A.R. potassium hydrogen phthalate, dry at 120°C for 2 h, and allow to cool in a desiccator over phosphorus pentoxide. Weigh out accurately approximately 0·6 g of salt, transfer to a 250 ml conical flask and dissolve in approximately 75 ml of distilled water. Add a few drops of phenolphthalein indicator and titrate with the sodium hydroxide solution. If w g of potassium hydrogen phthalate require V ml of sodium hydroxide solution for neutralisation.

$$\text{Normality of sodium hydroxide} = \frac{1000\,w}{204 \cdot 2\,V} \cdot \text{N}$$

$$= x\text{N}$$

To prepare 0·100N-sodium hydroxide, $100/x$ ml of the sodium hydroxide solution is diluted to 1 litre. Into a clean dry volumetric flask, run in from a burette $1000 - (100/x)$ ml of distilled water. Make up to the mark with the sodium hydroxide solution and shake well. Finally, titrate again against potassium hydrogen phthalate to ensure it is exactly decinormal.

Many other substances may be used to standardise the alkali (*see* Vogel [2]) but it is suggested that the choice is limited to A.R. quality potassium hydrogen phthalate, benzoic or succinic acid.

Although accurately standardised solutions of reagents are available commercially, it is desirable that students themselves should be competent to make up and standardise solutions required for volumetric analysis. Where the alkali solubility is employed as a routine test using large quantities of standard alkali, on grounds of both time and cost, it is convenient to make the sodium hydroxide solution from the solid and standardise against commercially available 0·100N hydrochloric acid solution.

(*b*) *Determination of Alkali Solubility* (*see* Smith and Harris [3] and I.W.T.O. [4]—A sample of wool (approximately 1·0 g dry) is spread out in an oven at 110°C for 1 h and weighed after transferring to a weighing bottle and allowing to cool in a desiccator for 10 min. If the sample is wet, it should be dried in the oven for the time indicated after it feels dry to the hand. Alternatively, samples may be weighed in equilibrium with atmospheric conditions and their dry weights calculated from the moisture content (in-duplicate) of a separate sample (*see* page xiii).

Measure 0·100N sodium hydroxide (100 ml) into a 250 ml tall beaker, and fit into the thermostat bath so that the temperature in the beaker is maintained at 65°C ± 0·5°C. (The water outside should be at about 68°C.) When the alkali reaches the required temperature, introduce the test sample of wool into the flask. If it is in the pre-spinning or yarn form, cut into lengths of approximately 1 cm, if fabric, cut into approximately 1 cm squares. Shake gently to ensure complete wetting (*see* Note 1).

After 60 min, transfer the contents of the flask to a weighed sintered glass filtering crucible. Wash any fibrous material from the flask into the crucible with distilled water. To avoid blocking the filter, apply gentle suction from the filter pump only as necessary. Wash the residue in the crucible six times with distilled water draining completely between each wash. Wash twice with dilute aqueous acetic acid (10 ml glacial acid per litre) and six more times with distilled water. Dry the crucible and contents for 3 h at 110°C, cool in a desiccator and weigh. The alkali solubility is the loss in weight of the wool expressed as a percentage of the original dry weight.

Duplicate determinations may be carried out on untreated wool and wool that has been chlorinated in acid solution (*see* page 94) or bleached with hydrogen peroxide (*see* Note 2). Although the wool may be in any form, it is convenient to carry out the chlorination and bleaching on patterns of fabric cut from the same piece as the untreated.

NOTES

1. It is not essential to carry out this process in a thermostat bath. The alkali may be contained in a beaker standing on a gauze and tripod. A small flame from a bunsen burner situated at the edge of the gauze will maintain the temperature of the alkali to within $\pm 0.5°C$. The temperature must, however, be checked every few minutes.

2. Wet out four 1 g wool patterns in water containing a few drops of 'Teepol'. Immerse in a solution consisting of water (95 ml) and 100 vol. hydrogen peroxide solution (4.0 ml) brought to pH 9–10 by the addition of a dilute solution of sodium hydroxide or preferably sodium silicate. The pH may be determined by test paper or universal indicator. Raise the temperature to 50°C, maintain at this temperature for 2 h, remove the source of heat and allow the wool to remain in the solution overnight. The patterns are rinsed well in water and dried. To determine their alkali solubility, the samples must be dried as previously indicated. If however, it is desired to retain reference samples of the bleached wool they should be dried at a lower temperature, for example room temperature.

REFERENCES

1 ALEXANDER, P., HUDSON, R. F. and EARLAND, C., *Wool—Its Chemistry and Physics.* Chapman & Hall, London, 293 (1963)
2 VOGEL, A. I., *Quantitative Inorganic Analysis.* Longmans, London, 243 (1961)
3 SMITH, A. and HARRIS, M., *Am. Dyes. Rep.,* **25**, 542 (1936)
4 I.W.T.O., *Technical Committee Report* 1960 and *J. Text. Inst.,* **52**, 678 (1961)

23 Effect of breaking disulphide cross-links in wool

APPARATUS

Thermostat bath, sintered glass filtering crucibles (No. 1 porosity), single-thread testing machine, preferably fitted with recorder for load-extension graphs.

MATERIALS

Glacial acetic acid, 100 vol. hydrogen peroxide, 0·100N-sodium hydroxide, scoured wool fabric and yarn.

THEORY

Peracetic acid breaks the disulphide cross-linkages in wool with the formation of cysteic acid residues. Although the following equation represents the main reaction, in addition intermediate oxidation products such as sulphones and sulphoxides may be formed also. When the cross-linkages are broken, the solubility of the wool in alkali is increased and the work necessary to stretch the fibre in the wet state is reduced.

EXPERIMENTAL PROCEDURE

(a) *Oxidation of Wool with Peracetic Acid* (*see* Note 1)—Prepare a stock solution of approximately 10 per cent peracetic acid by mixing

$$\overset{\displaystyle |}{\underset{\displaystyle |}{NH}} \qquad \overset{\displaystyle |}{\underset{\displaystyle |}{NH}}$$

$$CH \cdot CH_2 \cdot S \cdot S \cdot CH_2 \cdot CH \qquad \xrightarrow[\text{H}_2\text{O}]{5CH_3COOOH}$$

$$\overset{\displaystyle |}{CO} \qquad\qquad \overset{\displaystyle |}{CO} \qquad \text{Peracetic acid}$$

Cystine residue

$$CH \cdot CH_2 \cdot SO_3H + HO_3S \cdot CH_2 \cdot CH + 5CH_3COOH$$

$$\overset{\displaystyle NH}{} \qquad\qquad\qquad \overset{\displaystyle NH}{}$$

$$\overset{\displaystyle CO}{} \qquad\qquad\qquad \overset{\displaystyle CO}{}$$

Cysteic acid residues

glacial acetic acid (45 ml), 100 vol. hydrogen peroxide (55 ml) and concentrated sulphuric acid (0·6 ml) and allow to stand in a refrigerator (0–5°C) for at least 72 h to attain equilibrium.

$$CH_3COOH + H_2O_2 \rightleftharpoons CH_3COOOH + H_2O$$

The resulting solution is stable for several weeks if it is kept in the refrigerator. Immediately prior to use, prepare 200 ml of 2 per cent peracetic acid solution by diluting the stock solution (40 ml) with water (160 ml). Wet out seven 2 g patterns of wool fabric in water containing a few drops of 'Teepol', squeeze to remove excess water and immerse the patterns in the peracetic acid solution at room temperature. Remove samples at 2, 5, 10, 20, 30, 60 and 120 min. Wash for at least 15 min in several changes of water.

(*b*) *Effect of Breaking Disulphide Bonds on Alkali Solubility* (*see* Note 2)—Divide each oxidised pattern into two, dry at 110°C and weigh after cooling in a desiccator for 10 min. Determine the alkali solubility of the patterns (*see* page 72) in duplicate and plot a graph of alkali solubility versus time in the peracetic acid solution. The graph should include the alkali solubility of untreated (zero time) wool.

(*c*) *Effect of Breaking Disulphide Bonds on Tensile Properties*—

Prepare sixteen 1 g hanks of wool yarn, wet out and immerse fourteen in 2 per cent peracetic acid solution removing duplicate samples at the same intervals of time as (a). Wash off as before, and immerse all sixteen in 0·05M sodium bicarbonate (pH 8·5) prepared by dissolving 4–5 g of the solid in 1 litre of water. Allow eight samples (one untreated and seven treated for different times) to dry at room temperature. Determine the breaking load and extension at break of the dry and wet yarns treated with peracetic acid solution for different times. If the machine is fitted with a recorder, calculate the work to stretch the yarn a definite amount from the area between the graph, and a perpendicular dropped on to the extension axis. Tabulate the results and draw conclusions.

NOTES

1. In view of the fact that only dil. aq. peracetic acid solution is used, it is unnecessary to use alternative methods of preparation which result in the formation of more conc. products. Some of these methods, particularly those involving acetic anhydride, are hazardous.

2. The scope of these experiments may be extended by determining the cystine content (*see* page 52) of the wool oxidised with peracetic acid. Graphs should be plotted of the percentage loss of cystine versus (a) time of treatment with peracetic acid and (b) the alkali solubility of the wool. The changes in the mechanical properties of yarn may be correlated also with their cystine content.

REFERENCE
1 ALEXANDER, P., HUDSON, R. F. and EARLAND, C., *Wool—Its Chemistry and Physics*. Chapman & Hall, London, 266 and 291 (1963)

24 Cross-linking wool with formaldehyde

APPARATUS

No special apparatus is required apart from equipment for performing the alkali solubility test (*see* page 72).

MATERIALS

40 Per cent (w/v) formaldehyde solution (Formalin), 2 per cent aq. peracetic acid solution (*see* page 77), pH 6 buffer solution (*see* page 62) or sodium dihydrogen phosphate and sodium hydroxide, solutions for the alkali solubility test (*see* page 72) approximately 2 g samples of scoured wool fabric.

THEORY (*see* Fraenkel-Conrat *et al.* [1] and Alexander *et al.* [2])

Formaldehyde has been known as a cross-linking agent for a whole range of materials, particularly proteins and cellulose, since the beginning of the century. Although formaldehyde can react with many of the side-chains in wool, not all of these reactions give rise to cross-linkages.

Thus, although formaldehyde reacts with the free amino groups of lysine to give the methylol derivative:

$$W{-}NH_2 + HCHO \rightarrow W{-}NHCH_2OH$$

this cannot form stable linkages with amino groups in adjacent chains

$$W{-}NHCH_2OH + H_2N{-}W \rightarrow W{-}NHCH_2NH{-}W + H_2O$$

On the other hand, the methylol-lysine can give stable cross-links with amide groups

$$W-NHCH_2OH + H_2NOC-W \rightarrow W-NHCH_2HNOC-W$$
$$+ H_2O$$

Of the three recognised tests for cross-linking, increased work to stretch the fibre, reduced supercontraction and solubility in alkali (*see* Alexander [2]), the latter shows the greatest difference between untreated and formaldehyde treated wool.

EXPERIMENTAL PROCEDURE

Wet out two 2 g patterns of wool fabric and allow to stand in 2 per cent aq. peracetic acid solution (400 ml) for 10 min at room temperature with occasional agitation. Remove and wash very thoroughly with cold water.

If a pH 6 buffer solution is not available, dissolve 8·0 g sodium dihydrogen phosphate ($NaH_2PO_4 \cdot 2H_2O$) in water (1 litre) and add 2N-sodium hydroxide (20 g solid made up to 250 ml with water) until the pH is approximately 6. The latter may be determined with test-paper or a glass electrode (*see* page 16).

Prepare a 2 per cent solution of formaldehyde by mixing together formalin solution (10·0 ml), pH 6 buffer solution (20 ml) and making up to 200 ml with water. Gently boil one of the peracetic acid treated wool patterns in this for 1 h, placing a clock glass over the beaker to avoid the excessive escape of formaldehyde. (This operation is preferably performed in a fume cupboard and if solutions of formaldehyde come into contact with the skin they should be removed by washing in water.) Boil a 2 g sample of untreated wool in a similar solution of formaldehyde for 1 h. Wash both samples thoroughly with water, and after drying at 110°C determine the alkali solubility (*see* page 72) in duplicate of (a) untreated wool (b) wool oxidised with peracetic acid (c) wool cross-linked with formaldehyde and (d) wool oxidised with peracetic acid and then cross-linked with formaldehyde.

Tabulate and comment on the results.

REFERENCES
1 FRAENKEL-CONRAT, H. L., COOPER, M. and OLCOTT, H. S., *J. Am. chem. Soc.*, **67**, 950 (1945)
2 ALEXANDER, P., HUDSON, R. F. and EARLAND, C., *Wool—Its Chemistry and Physics*. Chapman & Hall, London, 324–331 (1963)

25 Formation of lanthionine linkages in wool

APPARATUS

No special apparatus is required.

MATERIALS

Borax (sodium tetraborate, $Na_2B_4O_7 \cdot 10H_2O$) patterns of scoured wool fabric (approximately 5 cm $\times$ 5 cm), preferably knitted.
 The above refer to Experimental Procedure (a) only.

THEORY

When wool is treated with alkali, the disulphide bonds of the cystine residues are replaced by thioether linkages with the elimination of sulphur.

 The thioether analogue of cystine is known as lanthionine and the modification of cystine in accordance with the equation given overleaf is described as the 'lanthionine reaction' (*see* Alexander *et al.* [1]). Compared with the disulphide cross-link, the lanthionine linkage is relatively unreactive and as a result many of the chemical reactions and physico-chemical properties of wool are changed considerably if it has received prior treatment with alkali.

EXPERIMENTAL PROCEDURE

(*a*) (*i*) *Formation of Lanthionine Linkages at* pH 13—Prepare approximately 0·1N sodium hydroxide by dissolving the solid

$$
\begin{array}{c}
| \\
\mathrm{NH} \\
| \\
\mathrm{CH-CH_2-S-S-CH_2-CH} \xrightarrow{\ \mathrm{OH}\ } \\
| \\
\mathrm{CO} \\
| \\
\end{array}
$$

Cystine residue

$$
\begin{array}{c}
| \\
\mathrm{NH} \\
| \\
\mathrm{CH-CH_2-S-CH_2-CH + S} \\
| \\
\mathrm{CO} \\
| \\
\end{array}
$$

Lanthionine residue

(4–5 g) in 1 litre of water. Immerse the wool patterns in this solution at room temperature for 24 h using a liquor to wool ratio of 100:1. It is advantageous to change the solution after 4 h to avoid the accumulation of sodium sulphide which reduces disulphide bonds. Wash the patterns for at least 15 min in running water and dry at room temperature.

(*a*) (*ii*) *Formation of Lanthionine Linkages at* pH 9—Prepare a 0·05M borax buffer solution of pH 9·2 by dissolving borax (19 g) in 1 litre of water. Using a water condenser, reflux the wool patterns in this solution for 2 h at a liquor to wool ratio of 100:1. Wash and dry as before.

(*b*) *Estimation of the Extent of Lanthionine Formation* (*see* Note)— The presence of lanthionine in wool hydrolysates may be shown by two-dimensional paper chromatography on Whatman No. 4 paper sheets 46 × 57 cm) using butanol—acetic acid—water solvent (60:15:25 v/v) in the first direction and phenol—water solvent (80:20 w/w) in the second (Figure 14). Since separation from cystine is not easy the small-scale equipment described on page 30 is not suitable. The chromatogram should be run for 24 h in each direction so that the solvent front runs off the paper, the edge of which is serrated to facilitate dripping. For reference purposes, samples of lanthionine are available commercially or it may be synthesised (*see* Atkinson *et al.* [1]). The wool may be analysed also for loss of cystine (*see* page 52).

(*c*) *Influence of Lanthionine Formation on the Physico-chemical Properties of Wool*

(i) *Alkali solubility* (*see* page 72). Determine the alkali solubility of wool in the untreated (scoured) state and after treatment with alkali at pH 9. This test is not valid on wool treated at pH 13.

(ii) *Yield of β-keratose fraction* (*see* page 85). The preparation of the alkali-insoluble fraction after wool has been oxidised with peracetic acid is carried out on untreated and alkali-treated (pH 9 or 13) wool. Method (b) (*see* page 87) is used, the product being treated with peracetic acid and ammonia twice prior to washing, drying and weighing. All ammoniacal solutions are discarded.

(iii) *Supercontraction in disulphide and hydrogen-bond breaking agents* (*see* page 64). Treat two 5 cm × 5 cm patterns of suitably marked wool fabric with alkali at pH 9 or 13. Determine the super-

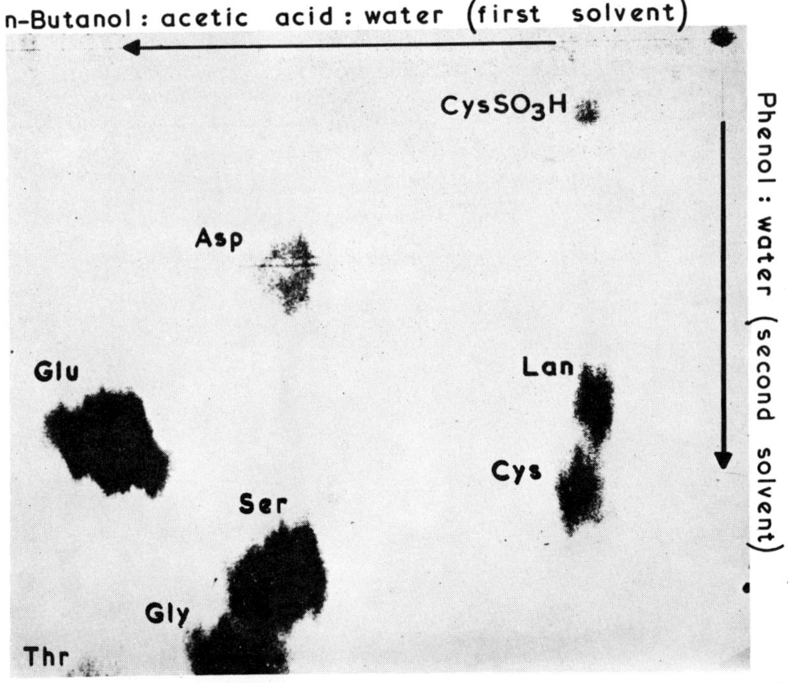

Figure 14.   Separation of lanthionine by two-way paper chromatography

contraction of an untreated and an alkali-treated pattern in hot 20 per cent sodium bisulphite solution. Determine also the supercontraction of similar patterns in 50 per cent (w/w) lithium bromide solution at 100°C for 1 min. Measure the contraction before and after washing in water. Tabulate the results and correlate them with the nature of the chemical modification in wool brought about by the alkali treatment and the mechanism of the supercontraction produced by the two reagents.

NOTE

The identification of lanthionine by paper chromatography is not recommended unless large-scale equipment is available and the student possesses some experience in this technique. Similarly, although the synthesis of lanthionine is not unduly difficult, it requires experience in preparative organic chemistry.

REFERENCES
1 ALEXANDER, P., HUDSON, R. F. and EARLAND, C., *Wool—Its Chemistry and Physics.* Chapman & Hall, London, 254 (1963)
2 ATKINSON, R. O., POPPELSDORF, F. and WILLIAMS, G., *J. chem. Soc.*, 580 (1953)

26 Isolation of the α-, β- and γ-keratose fractions from oxidised wool

APPARATUS

Mechanical stirrer, No. 2, porosity sintered glass crucible, apparatus for vacuum distillation, Soxhlet extractor, and (preferably) an electrically driven centrifuge.

MATERIALS

100 Vol. hydrogen peroxide, glacial acetic acid, ethyl alcohol (Industrial methylated spirit), ether, methyl alcohol, 64^s quality wool in grease, good quality soap flakes.

THEORY

The keratoses are sulphonic acid derivatives of keratin, prepared by oxidation of the disulphide bonds with aliphatic carboxylic per-acids, usually peracetic acid, followed by extraction with dilute ammonia. (For equation, see overleaf)

On acidification of the ammonia extract, a low-sulphur high-molecular weight fraction (α-keratose) is precipitated, leaving a high-sulphur low-molecular weight fraction (γ-keratose) in solution. The α-keratose fraction gives an x-ray diffraction pattern very similar to that of intact wool and must originate in the crystalline regions (microfibrils) of the fibre whereas the γ-fraction probably originates in an amorphous matrix. The residue insoluble in ammonia (β-keratose) is a morphological component of the fibre corresponding to the cortical cell membranes.

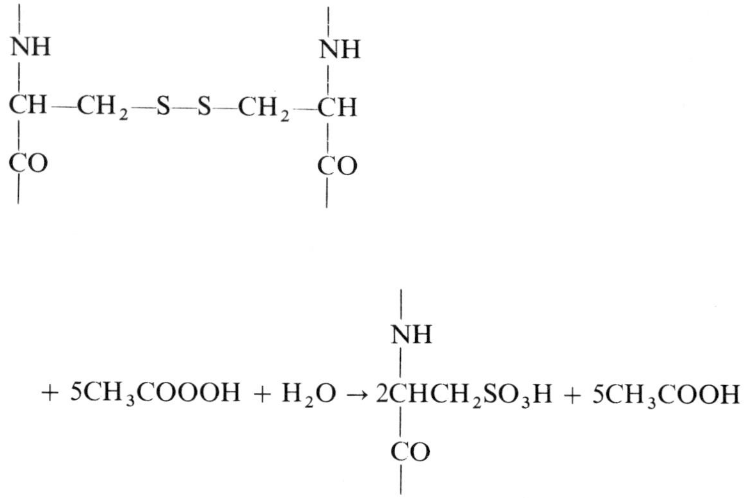

$$
\begin{array}{ccc}
\mid & & \mid \\
NH & & NH \\
\mid & & \mid \\
CH-CH_2-S-S-CH_2-CH \\
\mid & & \mid \\
CO & & CO \\
\mid & & \mid
\end{array}
$$

$$
+\ 5CH_3COOOH + H_2O \rightarrow 2\overset{\displaystyle \mid}{\underset{\displaystyle \mid}{CH}}CH_2SO_3H + 5CH_3COOH
$$

(with the central group bearing $-NH-$ above and $-CO-$ below the CH)

EXPERIMENTAL PROCEDURE

The complete flow sheet for the preparation of α-, β- and γ-keratoses is given in Figure 15 (*see* Note 1).

(*a*) *Purification and Oxidation of Wool*—Purified 64^s quality wool usually gives the following fractions based on the dry weight of the original wool—60 per cent α-keratose, 10 per cent β-keratose and 30 per cent γ-keratose. (These values may vary by 10 per cent.) If the wool has been modified in processing, however, such as in a hot alkaline scour, these quantities may be changed considerably (*see* page 81). It is essential, therefore, not to use commercially processed wool, but to purify wool in grease by the following procedure.

Make a scouring solution consisting of soap flakes (5 g) and 0·880 ammonia solution (1·0 ml) per litre. Agitate greasy wool (5 g) in about 300 ml of this solution for 10 min at 40°C. Repeat using two fresh portions of scouring solution. Wash free from soap in small portions of distilled water and air-dry. Remove burrs, etc., with tweezers and cut off any discoloured tips. Extract for 2 h with ether and then 2 h with methyl alcohol in a Soxhlet extractor and dry at room temperature or below 40°C.

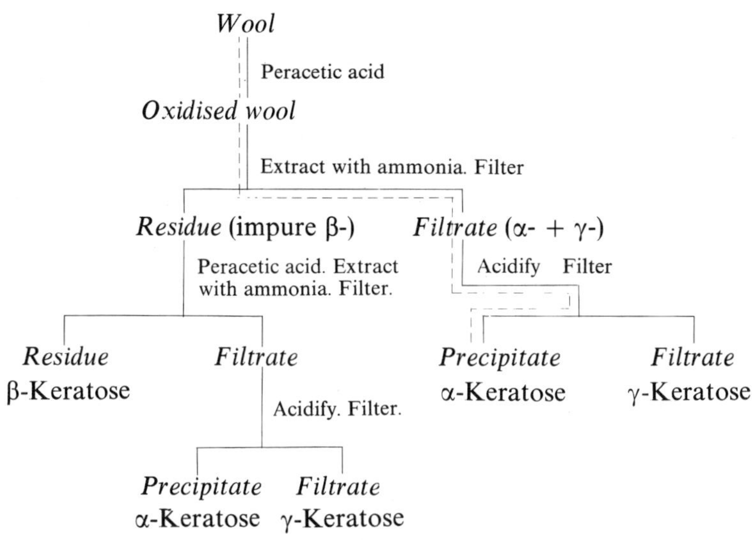

Figure 15.   Fractionation of oxidised wool

Take approximately 2·3 g of purified wool at normal regain. Loosely pack in a large weighing bottle (for example 100 × 25 mm) to ensure speedy drying and dry in an evacuated desiccator over phosphorus pentoxide for 3 days, checking the vacuum daily. Obtain the accurate dry weight of the wool which should be near to 2·0 g.

Treat with 2 per cent peracetic acid solution (50 ml), prepared as described on page 77, in a stoppered 100 ml flask with occasional shaking during 30 h at room temperature. Pour off and discard the peracetic acid (fume cupboard) and wash the oxidised wool with two portions of water (25 ml) in a sintered glass crucible (porosity No. 1 or 2).

(*b*) *Preparation of α- and β-Keratoses only*—Transfer the oxidised wool to a beaker containing 200 ml of approximately 0·2N-ammonium hydroxide (10 ml 0·880 ammonia per litre). Stir mechanically until the solid is dispersed completely (approx. 4 h). Filter using a funnel and filter paper (15 cm, No. 4). Wash the very gelatinous residue with water, saving the residue, filtrate and washings. Do not allow the residue to become dry on the paper or it will be difficult to remove and should it be necessary to leave overnight (the filtration is very slow), cover the funnel with a watch glass to prevent drying out.

The filtrate and washings from the ammonia extraction contain α- and γ-keratoses. Acidify to pH 1–2 with 2N-hydrochloric acid (25 ml) and allow to stand until the precipitate of α-keratose coagulates. Filter on a sintered glass crucible (porosity No. 2), rejecting the filtrate which contains γ-keratose and ammonium chloride and wash the residue of α-keratose with water until free of acid. Transfer to a weighed dish while still moist. If using suction for the filtration, do not allow the residue in the crucible to become too dry or it will be difficult to remove it quantitatively. Dry in an evacuated desiccator over silica gel and then over phosphorus pentoxide (3 days). Do not dry in the sintered glass crucible, because contraction of the α-keratose on drying disintegrates the sintered glass.

The gelatinous insoluble residue from the ammonia extraction consists of impure β-keratose and requires re-treatment with peracetic acid. Carefully remove from the paper while still wet, using a thin spatula or palette knife and transfer to a 100 ml flask. Add water (if necessary) and 10 per cent peracetic acid solution (10 ml) so that the final volume is about 50 ml (i.e the solution contains 2 per cent peracid). Allow to stand with occasional shaking for 30 h. Filter on a sintered glass crucible (No. 2), wash with water (50 ml) and transfer to 0·2N-ammonium hydroxide (200 ml). Stir until dispersed completely (about 2 h). Filter using a funnel and filter paper, wash the residue with water and retain the residue and also the filtrate and washings.

Acidify the filtrate with 2N-hydrochloric acid as before to obtain a further yield of α-keratose. Filter, wash and dry as described for the first crop of α-keratose, rejecting the filtrate. Weigh both crops of α-keratose and express the total weight as a percentage of the original dry weight of the wool.

The gelatinous residue from the second ammonia extraction consists of purified β-keratose. Remove from the paper, transfer to a glass dish and remove most of the moisture in an evacuated desiccator over silica gel, taking care to avoid frothing over. Finally dry *in vacuo* over phosphorus pentoxide for 3 days, weigh and calculate the yield as a percentage of the original dry weight of wool.
(*c*) *Preparation of* α-, β- *and* γ-*Keratoses*—Oxidise the wool with peracetic acid and disperse in ammonia solution as described under (*b*). Filter and acidify the filtrate containing α- and γ-keratoses with glacial acetic acid to pH 3–4 (5 ml) and allow to stand for the precipitate of α-keratose to coagulate. Filter as described under (*b*) and

retain the filtrate containing γ-keratose and ammonium acetate. Concentrate this to about 25 ml by distillation under reduced pressure at a temperature not exceeding 50°C (*see* page 120 and Figure 16).

Filter again and add dropwise to 250 ml of ethyl alcohol (industrial methylated spirit). Allow to stand for the precipitated γ-keratose to

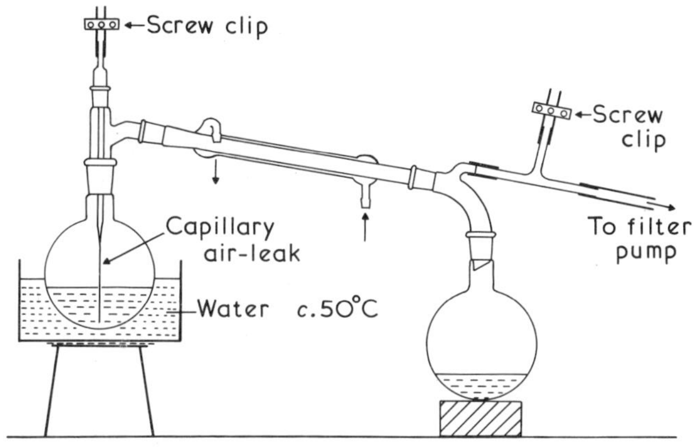

Figure 16.   Distillation under reduced pressure

coagulate. Ammonium acetate is soluble in alcohol and remains in solution. The precipitate of γ-keratose is difficult to recover by filtration and it is preferable to separate in a centrifuge. Wash with alcohol and transfer the slurry to a weighed dish. Dry under vacuum successively over silica gel and phosphorus pentoxide and determine the yield as described previously.

The residue from the ammonia extraction containing β-keratose is retreated as in (b) to determine the percentages of α- and β-keratoses. The second ammonia extraction may be acidified with hydrochloric acid to obtain the second crop of α-keratose as only a small amount of γ-keratose is present which may be rejected.

NOTES

1. In view of the length of time required for the complete experiment, the whole of the procedure given need not always be carried

out. Thus, α-keratose may be prepared by the route (- - - - -) shown, rejecting all residues and solutions not essential for its isolation.

2. The high molecular weight of α-keratose may be shown by the fact that it is possible to cast films and extrude filaments from this material. The freshly prepared α-keratose is moistened with a few drops of 0·880 ammonia to give a semi-solid dispersion. To prepare films, this is spread on a microscope slide and dried in a desiccator. Filaments may be prepared by extruding the same dispersion from a hypodermic syringe (No. 14 needle) into an aqueous solution containing 20 per cent Na_2SO_4 and 10 per cent H_2SO_4. Remove the filaments and without washing stand in aqueous formaldehyde solution (10 ml Formalin solution per 100 ml) for 2 h and air-dry without washing. If the equipment is available, compare the main x-ray diffraction spacings of the film and filaments with those of intact wool.

REFERENCE
1 ALEXANDER, P., HUDSON, R. F. and EARLAND, C., *Wool—Its Chemistry and Physics*. Chapman & Hall, London, 12 and 367 (1963)

27   Carbonising of wool

APPARATUS

Coarse sintered glass filter funnel.

MATERIALS

Sodium carbonate, potassium hydroxide, 'Teepol' detergent, 5 g samples of scoured fabric containing cotton and wool.

DISCUSSION

The object of carbonising is to remove vegetable matter, for example burrs, from virgin wool, or to recover wool from previously used material where it may be mixed with cotton or regenerated cellulose. The cellulose is degraded by acid and heat under conditions which cause little harm to the wool, and may be subsequently removed as dust.

In this experiment the commercial process is adapted to determine quantitatively the composition of a fabric containing wool and cotton.

EXPERIMENTAL PROCEDURE

(a) *Carbonising*—Weigh the sample of wool and cotton fabric which has been allowed to stand under normal laboratory conditions for a few days (*see* Note 1). Wet out in water containing a few drops of 'Teepol' for 5 min and squeeze to remove as much water as possible.

Immerse for 30 min in 100 ml of 4·5 per cent (w/v) sulphuric acid solution (25 ml concentrated acid per litre) with occasional stirring and squeezing. Remove the fabric and squeeze by hand as hard as possible.

Place the fabric on a watch glass and dry in an oven at 60°C followed by baking at 100–110°C for 30 min. Pulverise the material by hand and shake out the degraded cellulose. Rinse in warm water (*see* Note 2) and neutralise by immersing in 200 ml of 1 per cent sodium carbonate solution (or use 3 per cent $Na_2CO_3 \cdot 10H_2O$) at 40°C for 15 min. Wash with water and dry at 110°C.

When dry, transfer the residue of wool to a stoppered weighing bottle, allow to cool in a desiccator for 10 min and weigh.

$$\begin{aligned}
\text{If original weight of sample} &= w_1 \\
\text{Dry weight of residual wool} &= w_2 \\
\text{Wt. of wool under laboratory conditions} &= 1{\cdot}16w_2 \\
\text{Percentage wool in mixture} &= \frac{116w_2}{w_1}
\end{aligned}$$

(*b*) *Alternative Analytical Procedure*—As a check on the accuracy of the carbonising method, the fabric may be analysed by dissolving the wool in alkali and weighing the residual cotton.

Weigh a 5 g sample of the fabric under laboratory conditions, cut into squares approx. 1 cm side and boil for 10 min in 100 ml of 5 per cent aqueous potassium hydroxide solution. Filter the solution through a sintered glass funnel and wash the residual cotton with hot 5 per cent potassium hydroxide solution, followed by hot water, then approx. 1 per cent acetic acid solution and finally water. Dry at 110°C, transfer to a weighing bottle and weigh as previously.

$$\begin{aligned}
\text{If original weight of sample} &= w_1 \\
\text{Dry weight of residual cotton} &= w_2 \\
\text{Dry weight of cotton, corrected for solu-} & \\
\quad \text{bility in reagent} &= 1{\cdot}03w_2 \; (see \text{ Note 3}) \\
\text{Wt. of cotton under laboratory conditions} &= 1{\cdot}03 \times 1{\cdot}085w_2 \\
\text{Percentage cotton in mixture} &= \frac{112w_2}{w_1}
\end{aligned}$$

NOTES

1. Alternatively, all weighings may be made on completely dry

material and subsequently adjusted by means of 'official' regain values, for example using regain values of 8·5 per cent and 16·0 per cent for cotton and wool respectively. (*See* page xiii.)

If original dry weight of sample	$= x$
Dry weight of residual wool	$= y$
Then, dry weight of cotton	$= (x - y)$
Correct condition weight of wool	$= 1·16y$
Correct condition weight of cotton	$= 1·085(x - y)$
Percentage wool in mixture at correct condition	$= \dfrac{116y}{1·16y + 1·085(x - y)}$

2. This washing procedure may generally be carried out by decantation. If loose fibres separate, filtration through a sintered glass funnel is recommended.

3. Generally, between 2 and 5 per cent of the cotton dissolves in the potassium hydroxide solution. A mean value of 3 per cent has been taken.

28 Chlorination of wool fabric

APPARATUS

2·0 ml graduated pipette, microscope × 100.

MATERIALS

Sodium hypochlorite solution of known concentration (page 3), 5 per cent sodium bisulphite solution (or 5 g sodium metabisulphite in 100 ml water) 10 per cent ammonia solution (10 ml 0·880 ammonia + 90 ml water), borax, A.R. sodium thiosulphate, soap flakes, 'Teepol' detergent, 5 g patterns of scoured knitted wool fabric marked for shrinkage testing (page 62).

THEORY

Although the oldest method, the treatment of wool with chlorine in acid or alkaline solution is still used widely to render it non-felting. The degradation of the surface layer of the keratin results in the removal of the differential frictional effect from the fibre which is the chief cause of fibre migration. The reaction is complex, but involves chiefly the oxidative rupture of the cystine linkages between adjacent peptide chains with formation of sulphonic acid groups.

In acid solution:

$$W{-}S{-}S{-}W + 5Cl_2 + 6H_2O \rightarrow 2WSO_3H + 10H^+ + 10Cl^-$$

In alkaline solution:

$$W\text{—}S\text{—}S\text{—}W + 5OCl^- + H_2O \rightarrow 2WSO_3H + 5Cl^-$$

where W represents the protein chains of wool.

EXPERIMENTAL PROCEDURE

(*a*) *Chlorination in Solution of* pH 1—Weigh the wool pattern under normal atmospheric conditions to the nearest centigramme. Wet out in water containing a few drops of 'Teepol' for 5 min and rinse. Prepare 0·10N-sodium thiosulphate (*see* page 4) which need not be standardised and dilute to obtain 0·020N-sodium thiosulphate (page 97). Approximately N-sulphuric acid is prepared by adding 27 ml of conc. acid to 1 litre of water.

In a 600 ml beaker mix together water (350 ml) approximately N-sulphuric acid (40 ml) and sodium hypochlorite (V ml). The latter is calculated as follows:

$$V\ (\text{ml}) = \frac{\text{Percentage available Cl}_2 \text{ on wt. of wool} \times \text{wt. of wool (g)}}{\text{Available Cl}_2 \text{ in NaOCl soln. (g/100 ml)}}$$

In separate experiments, use 2·0 per cent and 4·0 per cent available chlorine on the weight of the wool.

Add the hypochlorite solution from the graduated pipette, stir and rapidly transfer 10·0 ml of the solution with a pipette to a 50 ml conical flask containing 0·2–0·3 g of potassium iodide. To avoid loss of chlorine, add the wool pattern immediately to the chlorine solution and stir gently for 15 min at room temperature. Withdraw further 10·0 ml samples at intervals of 1, 2, 5, 10 and 15 min and titrate iodometrically with the 0·020N sodium thiosulphate solution. Obtain an approximate value for the half-life of the reaction. After treatment, 'clear' the pattern of residual oxidising power by adding 5 per cent sodium bisulphite solution (approximately 10 ml) to the bath and allow to stand for 10 min. Rinse and then neutralise the samples by immersing in approximately 1 per cent ammonia solution (400 ml water + 40 ml 10 per cent ammonia) for 10 min at 40°C. The same neutralising bath can be used for both samples. Rinse and retain for further examination. Record the numbers of the patterns and the treatment received.

(*b*) *Chlorination in Solution of* pH 9 —Prepare a solution containing

19 g of borax per litre (0·05M). Repeat the chlorination procedure
(*a*), replacing the 40 ml of N-sulphuric acid by 50 ml of 0·05M borax.
Owing to the slower rate of reaction, stir the patterns in the bath
for 30 min and remove samples for iodometric titration at intervals
of 5 min. This titration must be performed in the presence of N-
sulphuric acid (5 ml). When clearing with sodium bisulphite, add
N-sulphuric acid (10 ml) to the bath. Neutralise and rinse as before.
(c) *The Milling Shrinkage of Chlorinated Wool Fabric*—Place the
four treated patterns together with one untreated in 1 per cent soap
solution at approximately 30°C for 5 min to wet out. Remove and
measure (page 63).

Hand-mill the samples together in the warm soap solution until the
untreated sample has shrunk about 20 per cent. Measure and
calculate the percentage area shrinkages of all the samples at this
stage. Continue the hand-milling until the untreated pattern has
shrunk about 40 per cent and again measure the shrinkages of all
the patterns, i.e. obtain shrinkage values for the treated samples
against controls of 20 and 40 per cent respectively. Tabulate the
results and draw conclusions.

After milling, examine patterns microscopically for scale damage
(page 102).

Chlorinated wool (not milled) may be analysed quantitatively for
cystine (page 52). and qualitatively for cysteic acid by electrophoresis
(page 39). Hydrolysates for the detection of cysteic acid may be
prepared as for the cystine determination.

REFERENCE

1 ALEXANDER, P., HUDSON, R. F. and EARLAND, C., *Wool—Its Chemistry and
Physics*. Chapman & Hall, London, 25–45 and 271 (1963)

29 A kinetic study of the reaction of wool with chlorosulphamic acid

APPARATUS

Mechanical stirrer, 2·0 ml graduated pipette, 10 ml burette, 10 ml pipette.

MATERIALS

Sulphamic acid, sodium chloride (B.P. is satisfactory) potassium iodide, sodium hypochlorite solution, sodium thiosulphate, starch, sodium metabisulphite, 'Teepol' detergent, knitted wool fabric.

EXPERIMENTAL PROCEDURE

Prepare solutions of approx. N-sulphuric acid (page 95), 0·10N-sodium thiosulphate (page 4) and determine the available chlorine in the sodium hypochlorite solution (page 3). Fifty millilitres of the sodium thiosulphate solution is made up to 250 ml with water to give an 0·020N solution.

The following is made up in a 600 ml beaker. Water (360 ml), 0·5M-sulphuric acid (40 ml), sodium chloride (40 g), sulphamic acid (0·5 g or 5 ml of a 10 per cent solution), and the calculated quantity of sodium hypochlorite solution.

Five gramme patterns of knitted fabric are prepared and marked for anti-felting testing (page 62). If not previously scoured, the patterns must be scoured, thoroughly washed and dried. They should be allowed to equilibrate under normal atmospheric conditions and weighed to the nearest 0·1 g.

The volume of sodium hypochlorite solution to be used (V) is given by the following:

$$V \text{ (ml)} = \frac{\text{Percentage available } Cl_2 \text{ on wt. of wool} \times \text{wt. of wool (g)}}{\text{Available } Cl_2 \text{ in NaOCl solution (g/100 ml)}}$$

In separate experiments, use 3·0, 4·0, and 5·0 per cent of available chlorine on the weight of wool.

Warm the bath to 40°C and transfer exactly 10·0 ml of solution by means of a pipette to a 50 ml conical flask. Add approximately 0·2–0·3 g of solid potassium iodide and titrate the liberated iodine with 0·020N-sodium thiosulphate from a 10 ml burette. A few drops of starch solution may be added near the end-point to assist its detection. Record the initial titration (a ml).

Wet out the pattern in water containing a few drops of 'Teepol' detergent for 5 min. Rinse, squeeze out and add the wool to the bath. Stir at approximately 100 rev/min continuously, keeping the temperature at 40·0 ± 0·5°C. This may be attained using a tripod and gauze and keeping the bunsen flame at the edge of the latter.

Remove 10·0 ml portions of solution at intervals of 5, 10, 15, 20, 25, 30 and 40 min, add potassium iodide and titrate against 0·020N-sodium thiosulphate as before. Record the titrations (T ml) ml.

After treatment, destroy the residual oxidising power of the samples by adding approximately 10 ml of 5 per cent sodium metabisulphite solution to the bath and allow to stand for 10 min. Rinse in water and neutralise the samples by immersing in approximately 1 per cent ammonia solution at 40°C for 10 min. Retain the samples for the next experiment.

CALCULATION OF RESULTS

Plot graphs of titration values versus time, t, for each run. It should be emphasised that in accordance with accepted nomenclature, the titration value T is represented by $(a - x)$ in Figure 17. Show that the reaction is of the first order since the time taken for the chlorine to half-exhaust, the half-life ($t_{\frac{1}{2}}$), is independent of the chlorine concentration. This is true, if the overall half-life of the individual runs be considered or the half-life at any time during a particular run be taken (*see* Figure 17a).

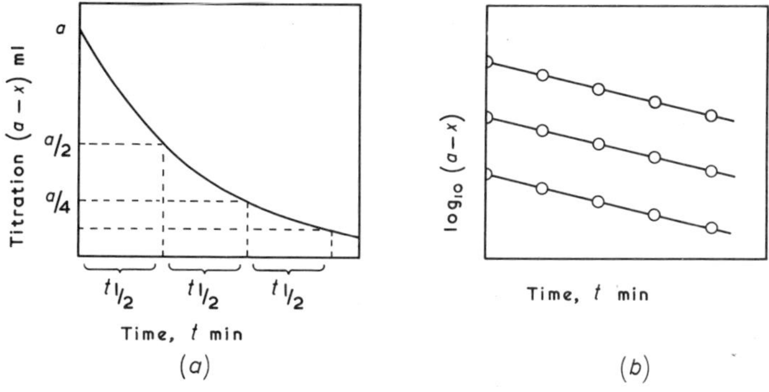

Figure 17.   Graphs of the kinetics of the reaction of chlorosulphamic acid with wool

From the half-life, calculate k, the rate constant.

$$k = \frac{2 \cdot 303 \log_{10} 2}{t_{\frac{1}{2}}} = \frac{0 \cdot 6931}{t_{\frac{1}{2}}} \; \text{min}^{-1}$$

Also plot graphs of $\log_{10}(a - x)$ versus t (Figure 17) and from the slopes, obtain values for k. In Figure 17b the intercept $= \log_{10} a$ and the slope $= -k/2 \cdot 303$.

THEORY

N-Chlorosulphamic acid is formed as follows:

$$NaOCl + H_2SO_4 \rightarrow HOCl + NaHSO_4$$

$$H \cdot SO_3 \cdot NH_2 + HOCl \rightarrow H \cdot SO_3 \cdot NHCl + H_2O$$

Sulphamic acid   N-Chlorosulphamic acid

The *overall* reaction with wool is essentially an oxidation of the cystine residues to sulphonic acid groups:

$$W{-}S{-}S{-}W + 5H \cdot SO_3 \cdot NHCl + 6H_2O \rightarrow 2WSO_3H$$
$$+ 5H \cdot SO_3 \cdot NH_2 + 5Cl^- + 5H^2$$

N-Chlorosulphamic acid in acid solution oxidises potassium iodide to free iodine.

H

$$2I^- + H \cdot SO_3 \cdot NHCl + H^+ \rightarrow H \cdot SO_3 \cdot NH_2 + I_2 + Cl^-$$

The progress of the reaction can, therefore, be conveniently followed by iodometric titration.

Using usual notations (*see* Latham [4]) the rate constant (k) for a first order reaction is given by

$$k = \frac{1}{t} \log_e \frac{a}{a-x} \tag{1}$$

$$k = \frac{2 \cdot 303}{t} \log_{10} \frac{a}{a-x} \tag{2}$$

Putting $a - x = a/2$

$$k = \frac{2 \cdot 303}{t_{\frac{1}{2}}} \log_{10} \frac{a}{a/2}$$

$$k = \frac{2 \cdot 303 \log_{10} 2}{t_{\frac{1}{2}}}$$

where

$$t_{\frac{1}{2}} = \text{time of half-life.}$$

Rearranging equation 2

$$\log_{10} \frac{a}{a-x} = \frac{k}{2 \cdot 303} t$$

$$\log_{10} a - \log_{10} (a - x) = \frac{k}{2 \cdot 303} t$$

$$\log_{10} (a - x) = \frac{-k}{2 \cdot 303} t + \log_{10} a$$

Comparing this with the straight line law, $y = mx + c$, a graph of $\log_{10} (a - x)$ against t will give a straight line of y intercept $= \log_{10} a$ and slope $\dfrac{-k}{2 \cdot 303}$.

REFERENCES

1 ALEXANDER, P., CARTER, D. and EARLAND, C., *J. Soc. Dyers. Cols.*, **66**, 538 (1950)
2 ALEXANDER, P., CARTER, D. and EARLAND, C., *J. Soc. Dyers. Cols.*, **67**, 17 (1951)
3 EARLAND, C., *J. Soc. Dyers. Cols.*, **71**, 89 (1955)
4 LATHAM, J. L., *Elementary Reaction Kinetics*, Butterworths, London **11**, 114 (1964)

30 Rendering wool non-felting with chloro-sulphamic acid

APPARATUS

See Experiment 29—microscope ($\times$ 100)

MATERIALS

See Experiment 29—soap flakes.

EXPERIMENTAL PROCEDURE

Chlorinate patterns of knitted wool fabric with 3, 4 and 5 per cent available chlorine, as chlorosulphamic acid, as described in the previous experiment. 'Anti-chlor' and neutralise also as described. In addition, carry out the reaction without sodium chloride in the solution. For each treatment record the half-life in minutes. The latter may be obtained from interpolation of the titration data.

Hand-mill the patterns in 1 per cent soap solution at approximately 30°C, with an untreated control, first taking the latter to approximately 20 per cent area shrinkage and then finally to 30 per cent (*see* pages 61 and 96). Owing to the number of patterns involved, it will be necessary to mill separately the patterns treated in the presence and absence of salt. Provided the controls in each set are milled to the same degree, this is not material. Tabulate the results as

Percentage available Cl_2 on fabric	Percentage NaCl in solution	$t_{\frac{1}{2}}$ (min)	Percentage area shrinkage	
0	0	—	20	30
3	0	—	—	—
4	0	—	—	—
5	0	—	—	—
3	10·0	—	—	—
4	10·0	—	—	—
5	10·0	—	—	—

The milling shrinkage figures in the first row are the means of the two untreated patterns. The figures in the last column of the table refer to the same patterns as the previous column, but they have been milled for a further period.

After milling, extract a few fibres from each pattern and wash thoroughly in distilled water to remove soap. Place on a microscope slide, add a drop of distilled water, cover with a slip and examine under the microscope ($\times 100$). Note the effect of the different treatments on the scale structure.

Wool patterns, after treatment with chlorsulphamic acid (unmilled), may be tested also for alkali solubility (page 72) and cystine content (page 52).

THEORY

There is no doubt that from the anti-felting viewpoint, the role of the sodium chloride is to modify the reaction between wool and chlorosulphamic acid so that degradation of the scale structure occurs.

Chemically, it was formerly assumed that chloramines reacted with wool via the slow liberation of chlorine or hypochlorous acid, and these reactions were simple chlorinations

$$RNHCl + H_2O \rightarrow RNH_2 + HOCl$$

or

$$RNHCl + HCl \rightarrow RNH_2 + Cl_2$$

It is now quite clear that the reactions are very complex and reference should be made to the literature cited on page 100.

31 Shrinkproofing wool by interfacial polymerisation of polyamides

APPARATUS

No special apparatus is required.

MATERIALS

Hexamethylene diamine ($H_2N[CH_2]_6NH_2$), adipoylchloride ($ClOC[CH_2]_4COCl$), sebacoyl chloride ($ClOC[CH_2]_8COCl$), anhydrous sodium carbonate, 'Teepol' detergent, carbon tetrachloride, scoured knitted wool fabric.

THEORY

The wool is treated with an aqueous solution of a diamine and then with a solution of a diacid chloride in an organic solvent. Formation of a polyamide, based on the Schotten–Baumann reaction, occurs at the interface of the two solutions, i.e. at the fibre surface.

$$H_2N[CH_2]_xNH_2 + ClOC[CH_2]_yCOCl$$

$$\downarrow$$

$$-HN[CH_2]_xNHCO[CH_2]_yCO- + 2HCl$$

In these experiments $x = 6$ and $y = 4$ or 8.

EXPERIMENTAL PROCEDURE

Cut the fabric into 5 g patterns and mark out for shrinkage testing (*see* page 62). Ensure that the samples are in equilibrium with atmospheric conditions and weigh four separately to the nearest centigramme (*see* Note 1). Take these patterns and wet out in water containing a few drops of 'Teepol' for 5 min. Rinse and squeeze out the excess water.

Immerse the four patterns with stirring and squeezing for 2 min in an aqueous solution (100 ml) containing hexamethylene diamine (8·8 g) and sodium carbonate (8 g). Squeeze to remove excess of solution.

Then immerse two of these patterns with stirring and squeezing for 2 min in carbon tetrachloride (50 ml) containing adipoyl chloride (1·0 ml) (*see* Note 2). Treat the remaining two patterns in an identical manner using sebacoyl chloride (2·0 ml) (*see* Note 2) in carbon tetrachloride (50 ml). Squeeze to remove excess solution, rinse the fabric thoroughly with water and dry at 110°C or at room temperature. Heating is not essential. Allow to equilibrate under laboratory conditions and reweigh to determine the polymer uptake.

It is not recommended that the solvent solution be applied to the wool prior to the diamine solution (*see* Whitfield *et al.* [1]).

Hand-mill the patterns (*see* page 61) in one per cent soap solution, recording their shrinkage after a control pattern has shrunk 10 per cent, 20 per cent and 30 per cent. All the treated patterns may be milled and mean shrinkage values for each determined or one pattern only need be used from each treatment and the other retained for reference.

After washing thoroughly with water to remove soap, and equilibrating under laboratory conditions, the weight of polymer remaining after milling may be determined.

NOTES

1. This treatment does not affect the water absorption properties of wool. If laboratory atmospheric conditions are liable to fluctuate excessively, determine the dry weight of the samples before and after polymer deposition. Calculate the weight of polymer deposited as a percentage on the dry wool and on 'conditioned' material, i.e. dry weight + standard regain (*see* page 92).

2. As a safety measure, these quantities should be measured from a burette or a graduated pipette, fitted with a rubber bulb.

REFERENCE
1 WHITFIELD, R. E., MILLER, L. A. and WASLEY, W. L., *Text Res. J.*, **31**, 704 (1961)

32 Properties and reactions of silk

APPARATUS

No special apparatus is required for experiments (a), (b) and (e).

MATERIALS

Bombyx mori silk in gum, scoured wool, good quality soap flakes, 90 per cent orthophosphoric acid, 90 per cent formic acid, acetic acid, anhydrous calcium chloride, sodium nitrite, nickel sulphate, cupric sulphate, cobaltous nitrate.

Note If experiments (c) are performed, apparatus and materials will be required to carry out paper or thin-layer chromatography (*see* pages 30 and 33) and tyrosine analyses (*see* page 56).

DISCUSSION

The silk used in the textile industry is the filament produced by a caterpillar to form a cocoon to protect the insect during its metamorphosis from larva to moth. Although thousands of these insects (*Lepidoptera*) produce silks, the commercial material is obtained mainly from the cultivated species *Bombxy mori* and to a lesser extent from the semi-wild species *Antheraea pernyi* (Tussah silk). The thread as extruded by the worm consists of two filaments of fibroin cemented together by a gum known as sericin. The latter is soluble in aqueous solutions, for example soap, and is removed during the processing of the silk.

All silks are protein in nature, yielding α-amino acids on hydrolysis.

Although different silks may vary greatly in amino acid composition, none are cross-linked like wool due to the virtual absence of cystine residues. Thus, unlike wool, silks are soluble in a number of non-destructive solvents. The term 'silk', without qualification, refers to *Bombyx mori* fibroin, 90 per cent of which by weight may be accounted for by the residues of glycine, alanine, serine and tyrosine. (*See* Peters [1].)

EXPERIMENTAL PROCEDURE

(*a*) *Determination of sericin and fibroin*—Take *Bombyx mori* silk in gum (approximately 4 g), dry in an oven for 1 h at 110°C and weigh accurately. Boil for 30 min in a solution consisting of soap flakes (5 g) and water (500 ml). Pour off the solution and boil for a further 30 min in fresh soap solution. Pour off and wash the residue of fibroin with five separate quantities (100 ml each) of distilled water to remove soap. Remove excess water with filter paper and dry for 90 min at 110°C prior to weighing. The percentage weight loss corresponds to the sericin originally present. Retain the purified fibroin for the following experiments.

(*b*) *Effect of solvents* (Earland and Raven [2])—Prepare calcium chloride–formic acid solvent by dissolving anhydrous (fused) calcium chloride (5 g) in 90 per cent formic acid (50 ml) (CAUTION—burns skin). If the solution is turbid, filter through a paper. Examine the effect of this solvent and 90 per cent orthophosphoric acid (approximately 10 ml) on small quantities of degummed silk and scoured wool (approximately 0·2 g). Pour the solvent off from the residue, if any, into 250 ml flasks and to each of the four solutions add cold water (100 ml) mixing *gently*. Note the effect. Now shake the flasks very vigorously and again observe the result.

(*c*) *Amino acid composition*—Hydrolysates may be prepared by refluxing the silk (1 g) with 5N-hydrochloric acid (25 ml) for 8 h and removing the excess acid by evaporation in a vacuum desiccator over solid sodium hydroxide. The residue is analysed qualitatively for amino acids by paper (*see* page 30) or thin-layer chromatography (*see* page 33).

Silk may be analysed quantitatively for tyrosine by the method given on page 56 hydrolysing an accurately known weight of the dried material either by the sealed tube method (*see* page 54) or the open reflux procedure described previously.

(*d*) *Preparation of metal nitroso-complexes*—The action of nitrous acid and certain metallic salts on tyrosine-containing protein fibres is to produce coloured metal complexes. Since these are chemically bound to the fibre, the colours are completely fast to washing (*see* Nilssen [3]). Complexes of silk may be prepared as follows:

Dissolve without heating sodium nitrite (1–2 g) in a mixture of water (90 ml) and acetic acid (10 ml). Take three boiling tubes and place about 30 ml of this solution in each. To each add silk (approximately 0·1 g) followed by the appropriate metal salt (approximately 0·1 g). Warm to about 60°C with shaking and allow to stand for 10 min to develop the colour. Remove the silk, wash in water and dry. Note the colours produced by nickel sulphate, copper sulphate and cobaltous nitrate.

The same method may be used to prepare similar complexes with wool keratin.

REFERENCES
1 PETERS, R. H., *Textile Chemistry*, vol. I, Elsevier, Amsterdam, 302 (1963)
2 EARLAND, C. and RAVEN, D. J., *Nature. Lond.*, **174**, 461 (1954)
3 NILSSEN, B., *Symp. Fibrous Proteins* (Soc. Dy. Col. Bradford), 142 (1946)

PART 3   Experiments on Carbohydrates

33 Preparation of α- and β-glucose penta acetates

APPARATUS

Melting-point apparatus.

MATERIALS

Acetic anhydride, acetone, methanol or industrial methylated spirit, glucose, anhydrous zinc chloride, anhydrous sodium acetate, $1 \cdot 0_N$-sodium hydroxide of exactly known normality, $1 \cdot 0_N$-hydrochloric acid (approximate normality only required) (*see* page 144).

THEORY

On heating glucose with an excess of acetic anhydride in the presence of a catalyst, all five hydroxyl groups are acetylated. The resulting penta-acetate can exist as two stable stereo isomers corresponding to the α- and β-forms of glucose. With zinc chloride as catalyst, the α-form is the main product and with sodium acetate the β-form predominates.

It should be noted that isomers can arise during this reaction only from groups attached to carbon atom 1 since the ring can break. and reform at this point, whereas the spatial arrangement of the groups attached to the other carbon atoms is fixed (*see* McIlroy [1]).

This experiment is confirmatory evidence for the existence of two forms of D-glucose since although the two derivatives have

The reaction schemes show glucose reacting with $(CH_3CO)_2O$ / $ZnCl_2$ to form α-Glucose penta-acetate, and with $(CH_3CO)_2O$ + $CH_3CO\cdot ONa$ to form β-Glucose penta-acetate.

widely differing melting points, analysis shows they both correspond to penta-acetates of glucose.

EXPERIMENTAL PROCEDURE

(a) *Preparation of α-glucose penta-acetate*—Place anhydrous zinc chloride (1·0 g) and acetic anhydride (25 ml) in a 250 ml round bottomed flask, attach a reflux condenser and heat on a boiling water bath for 5–10 min. Add powdered glucose (5 g) slowly, shaking the mixture gently during the addition, to control the vigorous reaction which occurs. Heat the flask on a boiling water bath for 1 h and pour the reaction mixture into 250 ml of water containing crushed ice. Stir the mixture and cool in ice for 30 min. The penta-acetate will first separate out as an oil but this will solidify on stirring. Filter at the pump, wash with a little cold water and recrystallise from 40 per cent (v/v) aqueous methanol or 40 per cent industrial methylated spirit until the melting point is constant at 110–111°C.

Two recrystallisations are generally sufficient. Dissolve the product in the minimum quantity of hot solvent and for the first recrystallisation, stir for a few minutes with about 1 g of decolorising charcoal prior to filtration through a fluted paper (*see* Vogel [2]).

Record the melting point and yield and calculate the latter as a percentage of the theoretical.

(*b*) *Preparation of β-glucose penta-acetate*—Mix together powdered anhydrous sodium acetate (4 g) and dry glucose (5 g) and place the mixture in a 250 ml round bottomed flask. Add acetic anhydride (25 ml), attach a reflux condenser, and with occasional shaking, heat on a boiling water bath until the solution is clear (approximately 30 min). Continue the heating for a further 2 h. Pour the reaction mixture into 250 ml of water containing crushed ice in a beaker. Break up the lumps with a glass rod and allow to stand with occasional stirring for 1 h. Filter the crystals at the pump, wash well with cold water and recrystallise twice as for the α-penta-acetate. Record the melting point (131–132°C) and yield.

(*c*) *Conversion of β-glucose penta-acetate into the α-form*—Add anhydrous zinc chloride (0·5 g) rapidly to acetic anhydride (25 ml) in a 250 ml round bottomed flask. Attach a reflux condenser, heat on a boiling water-bath for 5–10 min to dissolve the solid, add β-glucose penta-acetate (5 g) and heat on the water bath for a further 30 min. Pour the hot solution into 250 ml of water containing crushed ice and stir vigorously to induce crystallisation of the oil. Filter and recrystallise from 40 per cent aq. methanol or 40 per cent aq. methylated spirit as previously. Record the yield and melting point. The latter should be determined also on an intimate mixture of the product and the α-glucose penta-acetate prepared by method (a) (*see* Vogel [2]).

(*d*) *Determination of acetyl values*—Place about 2 g of the glucose penta-acetate in an unstoppered weighing bottle and leave in an evacuated desiccator over phosphorus pentoxide for at least 24 h. Weigh and transfer the contents to a 500 ml conical flask. Dissolve in 150 ml of acetone, add 30 ml N-NaOH (of exactly known normality) and 20 ml of water for hydrolysis. To a blank of 150 ml of acetone, add exactly the same volume of N-NaOH and water. Titrate this blank with N-HCl using phenolphthalein as indicator. (The exact normality of the HCl is not required.) After allowing to stand for a few minutes to complete the hydrolysis, add exactly the same volume of N-HCl to the determination. Titrate the determination with N-NaOH using phenolphthalein. (This extra acidity of the determination is due to the acetic acid liberated.)

$$\text{Acetyl value} = \frac{6 \times \text{Volume of NaOH (ml)} \times \text{Normality of NaOH}}{\text{Weight of sample (g)}}$$

Calculate the theoretical acetyl value of glucose penta-acetate. Acetyl value is the number of grammes of acetic acid obtained by saponification of 100 g of the acetate ester.

To gain experience in this determination, the acetyl value of the α-glucose penta-acetate prepared by method (a) may be determined in duplicate. The limited material available will, however, permit of only single determinations on the acetates prepared by methods (b) and (c).

REFERENCES
1 McIlroy, R. J., *Introduction to Carbohydrate Chemistry*. Butterworths, London, 22 (1967)
2 Vogel, A. I., *Practical Organic Chemistry*, Longmans, London, 34, 127 (1956)

34 Preparation of cellobiose octa-acetate from cellulose

APPARATUS

Thermostat bath, Buchner or sintered glass funnel, melting-point apparatus.

MATERIALS

Purified wood pulp or cotton linters (*see* page 141), glacial acetic acid, acetic anhydride, ethyl alcohol (industrial methylated spirit) acetone, N-NaOH solution (normality known), N-HCl solution (approximate normality required).

THEORY

The reaction is a combined hydrolysis of cellulose to cellobiose and the acetylation of the latter to the octa-acetate (for equation see overleaf).

The theoretical yield is calculated from

$$C_{12}H_{20}O_{10} \rightarrow C_{12}H_{14}O_3(OCOCH_3)_8$$

Cellobiose residue   Cellobiose octa-acetate

and the acetyl value from

$$C_{12}H_{14}O_3(OCOCH_3)_8 + 8H_2O \rightarrow C_{12}H_{22}O_{11} + 8CH_3COOH$$

Cellobiose

115

I

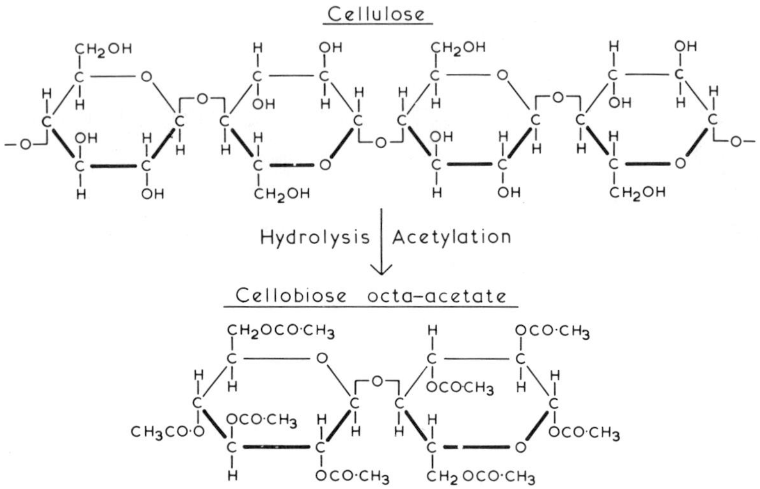

Cellulose

Hydrolysis | Acetylation

Cellobiose octa–acetate

EXPERIMENTAL PROCEDURE

Place the air-dry purified wood pulp or cotton linters (10 g) in a mixture of glacial acetic acid (40 ml), acetic anhydride (40 ml) and conc. sulphuric acid (4·1 ml) contained in a 100 ml conical flask. Place the stoppered flask in a thermostat bath at 30°C for 14 days.

Filter off the cellobiose octa-acetate which separates, using a sintered glass funnel. Add water (350 ml) to the filtrate and allow to stand in a refrigerator to obtain a further crop of crude material.

Combine both crops and heat under reflux with industrial methylated spirit (300 ml) on a hot water bath or electric heating mantle. When no further material dissolves (the alcohol present is sufficient to dissolve all the octa-acetate), filter hot at the water pump and allow the filtrate to cool, preferably standing in the refrigerator overnight. Filter off the recrystallised product using a Buchner funnel and filter paper or a sintered glass funnel, transfer to a weighed dish and dry in a vacuum desiccator over silica gel. Determine the yield and melting point of the product. The melting point of the pure ester should be 228–229°C with darkening at about 220°C. If the melting point is below 225°C, repeat the re-crystallisation adding about 1 g of decolorising charcoal to the alcohol solvent (filter through paper but not sintered glass).

Determine the acetyl value (*see* page 113), dissolving about 1·5 g (accurately weighed) of dry octa-acetate in 150 ml of acetone.

Retain the remainder of the preparation for conversion to cellobiose
(*see* page 118).

Record (a) the yield in grammes (b) the yield as a percentage of
the theoretical (c) the melting point (d) the acetyl value and (e) the
theoretical acetyl value of the product.

REFERENCE
1 Hess, K. and Friese, H., *Annalen*, **456**, 49 (1927)

35 Preparation of cellobiose from cellobiose octa-acetate

APPARATUS

Three-necked flask (500 ml), mechanical stirrer, ion-exchange column, Shandon Unikit paper chromatography apparatus (*see* page 26), apparatus for distillation under reduced pressure (*see* page 89).

MATERIALS

Cellobiose octa-acetate (*see* page 115), sodium hydroxide, ethyl alcohol (industrial methylated spirit), acetic acid, ethyl acetate, acetone, pyridine, aniline-phthalate reagent (*see* Note 1), Dowex 50W-X8 (20–50 U.S. mesh) or Zeo-Karb 225, 8 per cent cross-linked (14–52 B.S.S. mesh) ion-exchange resin, Whatman No. 1 chromatography paper, nitrogen cylinder, decolorizing charcoal, glucose, cellobiose, phenylhydrazine hydrochloride, sodium acetate.

THEORY

On alkaline hydrolysis (saponification) cellobiose octa-acetate yields the free disaccharide.

$$C_{12}H_{14}O_3(OCOCH_3)_8 + 8NaOH \rightarrow C_{12}H_{14}O_3(OH)_8$$

Cellobiose   Cellobiose
octa-acetate

$$+ 8CH_3COONa$$

Glucose may be produced as a by-product due to hydrolysis to the monosaccharide.

$$C_{12}H_{22}O_{11} + H_2O \rightarrow 2C_6H_{12}O_6$$

Cellobiose   Glucose

Cellobiose and sodium acetate are both soluble in water and removal of sodium ions is carried out by means of a cation-exchange resin.

$$RSO_3^- H^+ + CH_3COO^- Na^+ \rightleftharpoons RSO_3^- Na^+ + CH_3COOH$$

Resin in   Resin in
acid form   salt form

EXPERIMENTAL PROCEDURE

(a) *Saponification of cellobiose octa-acetate*—In a three-necked 500 ml flask place cellobiose octa-acetate (5 g), sodium hydroxide (4 g) dissolved in water (100 ml) and ethyl alcohol (100 ml). Through the centre neck insert a mechanically-driven stirrer with a gland to seal the opening. Into one of the side necks insert a rubber stopper and delivery tube connected to a nitrogen cylinder. This delivery tube must dip below the surface of the liquid but must be clear of the stirrer. Through the remaining neck insert a rubber stopper and outlet tube. If preferred, rubber stoppers may be replaced by Quickfit screw thread adapters. The outlet tube should terminate just below the neck inside the flask and dip into a beaker of water outside the flask (*see* Figure 18). Stir for 60 min at room temperature passing nitrogen through the flask to prevent alkaline atmospheric oxidation. A steady stream of bubbles from the outlet tube indicates that the apparatus is air-tight and nitrogen is passing at a satisfactory rate. Add acetic acid (8 ml) and filter. Concentrate the solution of cellobiose and sodium acetate to about 25 ml by distillation under reduced pressure (*see* Figure 16). The capillary leak is to minimise bumping and frothing into the condenser. Should this however, occur, the vacuum should be released by opening the clip nearest to the pump.

(b) *Removal of sodium ions*—The ion-exchange column of length 30–40 cm and diameter approximately 2 cm is set up as shown in Figure 19 and contains cross-linked polystyrene sulphonic acid cation-exchange resin (Dowex 50 or Zeo-Karb 225) which must be

119

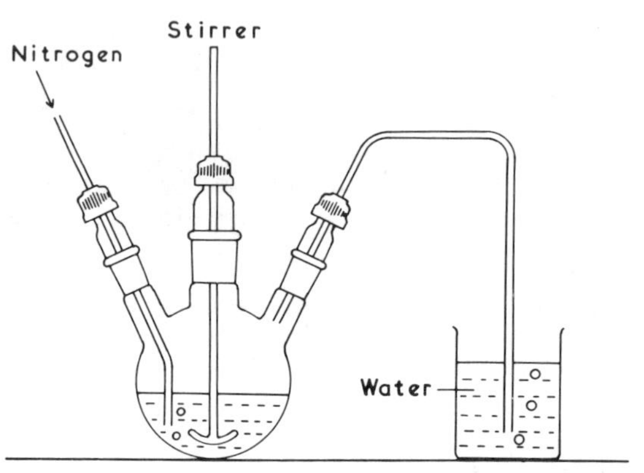

Figure 18.   Apparatus for saponification of cellobiose octa-acetate

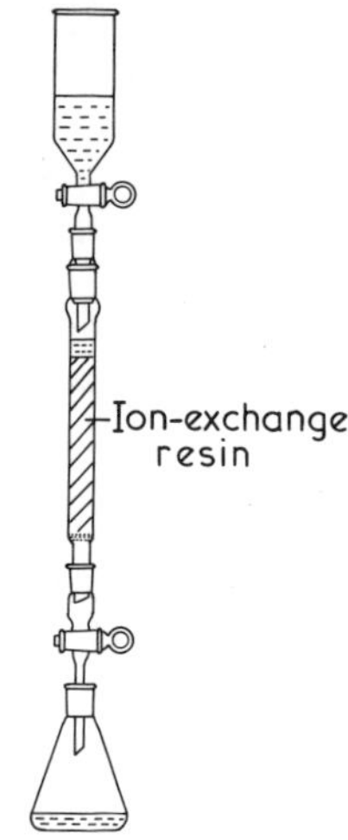

*Figure 19.
Ion-exchange column*

in the acid (hydrogen) form (*see* Note 2). Pass the solution through the resin over a period of about 20 min. The sodium ions are retained by the column and cellobiose and acetic acid pass through. Do not allow the liquid level to fall below the resin layer or air bubbles will become trapped in the column and reduce its efficiency.

Collect the effluent from the column in about 10 ml portions and after all the solution has passed into the column, pass distilled water (50 ml) through and continue collecting fractions. Determine which fractions contain cellobiose by subjecting them to paper chromatography together with a known sample of cellobiose. For practice, other sugars may be run on the same paper, for example glucose, lactose, xylose, etc. Use a 25 cm square of Whatman No. 1 paper in a Shandon Unikit tank (*see* page 26) with ethyl acetate–pyridine–water (55:25:20 by volume) as solvent and run for 2 h. Mark the position of the solvent front and dry the paper, suspended from stainless steel clips, in a fume cupboard. Spray with aniline-phthalate reagent and then heat the paper in an oven at 100–110°C for a few minutes to locate the positions of the sugars. Measure the R_F values of the spots obtained (*see* page 26).

Combine the fractions containing the cellobiose and warm them with decolorising charcoal (1 g) to remove any yellow colour. Filter and transfer the filtrate to the minimum size beaker and evaporate to dryness in a vacuum desiccator over solid sodium hydroxide. Add water (1·0 ml) and heat cautiously to boiling. Add water dropwise to completely effect solution, cool and add an equal volume of acetone. Allow to stand in the refrigerator to complete the crystallisation. Filter off the cellobiose and dry at 110°C. Determine the yield and calculate as a percentage of the theoretical. Record the melting-point (230–235°C with decomposition) and repeat the chromatography on a few milligrammes of the final product dissolved in water. A possible impurity is glucose, so run some glucose on the chromatogram in addition to an authentic sample of cellobiose. Finally prepare the osazone.

(*c*) *Preparation of cellobiose osazone* (McIlroy [1])—Place cellobiose (0·2 g), A.R. phenylhydrazine hydrochloride (0·4 g) hydrated sodium acetate (0·6 g) and water (4 ml) in a test-tube. Stopper lightly and stand upright in a beaker of boiling water for 15 min. Cool and filter off the precipitated osazone on a small filter paper and wash with a little cold water. Recrystallise from hot water and examine the crystalline form under the microscope. It may be necessary to repeat the recrystallisation to obtain good results. Photographs of osazone crystals are given by Vogel [2]. Dry the crystals at 110°C and determine their melting point (198°C). As many osazones have similar melting points which may be indefinite, their crystalline form is a more positive guide to their identification than their melting point.

NOTES

1. The aniline-phthalate reagent may be purchased as an aerosol spray. It is much cheaper, however, to dissolve phthalic acid (1·6 g) and aniline (0·9 ml) in n-butanol saturated with water (100 ml) and apply this from an ordinary chromatography spray.

2. To convert the resin from the salt form to the acid (hydrogen) form, pass 2N-hydrochloric acid (80 ml conc. acid + 400 ml water) slowly (*c.* 30 min) through the column. Wash with distilled water until the effluent is no longer acid. The capacity of the resin is about 5 m.equiv./g, i.e. 100 g of resin will absorb the sodium ions from 41 g of sodium acetate.

REFERENCES
1 McILROY, R. J., *Introduction to Carbohydrate Chemistry*. Butterworths, London, 9 (1967)
2 VOGEL, A. I., *Practical Organic Chemistry*. Longmans, London, 455 (1956)

36 Determination of α-, β- and γ-cellulose

APPARATUS

No. 2 porosity 3 in sintered glass funnel, 2 litre filter flask.

MATERIALS

Acetic acid, 90 per cent orthophosphoric acid, ammonium ferrous sulphate (A.R.), potassium dichromate (A.R.), sodium diphenyl-aminesulphonate indicator (0·2 per cent solution), wood-pulp cellulose.

DISCUSSION

The terms α-, β- and γ-cellulose were introduced by Cross and Bevan in 1904 and refer to the fractionation of cellulose by the action of conc. sodium hydroxide solution. The insoluble fraction is called α-cellulose. The material soluble in the alkali but which is precipitated on acidification is known as β-cellulose, whilst the fraction which remains in solution is γ-cellulose. It should be emphasised that the terms α, β and γ have no chemical significance and α-cellulose, which may be regarded as chemically pure cellulose, is composed of cellobiose residues in which the glucose residues are in the β-configuration.

In the method described, which is used widely in the rayon industry, the α-cellulose is determined gravimetrically and the β- and γ-celluloses are determined volumetrically, making use of the fact that cellulose is oxidized quantitatively by acidified potassium dichromate solution

$$C_6H_{10}O_5 + 12\,O \rightarrow 6CO_2 + 5H_2O$$

$\therefore$ 24 litres N-$K_2Cr_2O_7 \equiv 162$ g cellulose

or x ml of y N-$K_2Cr_2O_7 \equiv \dfrac{162xy}{24\,000}$ g cellulose

$$= 0{\cdot}00675xy \text{ g cellulose}$$

The equivalent weight of potassium dichromate is one-sixth of its molecular weight (*see* page 135).

EXPERIMENTAL PROCEDURE

(*a*) *Gravimetric determination of* α-*cellulose*—Prepare the following:

(i) A 17·8 per cent (w/w) caustic soda solution (18 g NaOH + 82 ml distilled water) adjusted to 20°C.
(ii) One litre of distilled water adjusted to 20°C.
(iii) One litre of hot distilled water ready to bring to the boil.
(iv) Ten per cent (w/v) acetic acid (10 ml glacial acid + 90 ml distilled water).

Weigh accurately two portions of 10 g of wood-pulp cut into pieces approximately 1 cm square. Place one sample in an unstoppered weighing bottle and dry in an oven at 110°C for 3 h. Allow to cool in a desiccator, replace the stopper, and re-weigh to determine the moisture content.

Place the other sample in a 400 ml glass mortar and add 17·8 per cent sodium hydroxide solution (50 ml) at exactly 20°C. Knead for 5 min with a glass rod flattened at one end. Then allow to stand for a further 20 min at 20°C. Loosen the pulp and add distilled water (150 ml) at 20°C. Stir rapidly until the mass is homogeneous.

Five minutes after the addition of the water, i.e. exactly 30 min from the time of the addition of the sodium hydroxide solution, pour the contents of the mortar into a No. 2 porosity sintered funnel, draining by means of a filter pump into a clean 2 litre filter flask, taking care not to draw air through the pulp. Wash with the remainder (850 ml) of the water at 20°C over a period of 5 min, again not sucking air through the mass. Reserve the filtrate for the determination of β- and γ-celluloses (*see* Section (b)).

Change the filter flask and pour 10 per cent acetic acid (100 ml) over the pulp and allow to percolate through slowly. Finally wash with boiling water (1 litre) with stirring. Press the cellulose well down and dry the funnel overnight at 110°C. Then transfer to a large weighing bottle and dry for a further 4 h at 110°C. Allow to cool in a desiccator, replace the stopper, and weigh.

$$\text{Percentage } \alpha\text{-cellulose} = \frac{\text{Wt. of } \alpha\text{-cellulose residue}}{\text{Dry weight of original pulp}} \times 100$$

Duplicate analyses should agree within ±0·2 per cent. If only one oven is available, funnels containing wet pulp should not be placed in it before the samples being dried for moisture content have been removed.

(*b*) *Volumetric determination of* β- *and* γ-*celluloses*—Prepare the following:

(i) Make the filtrate containing β- and γ-celluloses up to 1 litre with distilled water in a volumetric flask.

(ii) 2N-Sulphuric acid by adding 56 ml of conc. acid to 500 ml water. Cool and make up to 1 litre in a volumetric flask.

(iii) Weigh out ammonium ferrous sulphate (approximately 60 g) and dissolve in 180 ml $2\text{N-H}_2\text{SO}_4$. Make up to 1 litre with distilled water in a volumetric flask. (This solution is approximately 0·15N.)

(iv) Weigh out accurately exactly 21·000 g of dry A.R. potassium dichromate. Dissolve in distilled water and make up to 250 ml in a volumetric flask. (This solution is 1·713N.)

Transfer 100 ml of the β + γ filtrate to a 500 ml conical flask and then pipette 25 ml of the potassium dichromate solution into this. Rotate the flask whilst adding conc. sulphuric acid (100 ml) with care. Allow the hot solution to stand for 10 min, then cool and make up to 250 ml in a volumetric flask. Shake well.

Pipette 25 ml of this solution into a 250 ml conical flask and dilute to 100–150 ml. Add phosphoric acid (10 ml) and sodium diphenyl-aminesulphonate indicator (8 drops). Titrate with the ammonium ferrous sulphate solution from the pinky-violet colour to a clear green.

Dilute 25 ml of the potassium dichromate solution from a pipette to 250 ml in a volumetric flask, shake well and titrate 25 ml with the ammonium ferrous sulphate solution as previously. Calculate the normality of the ammonium ferrous sulphate solution (y N) from this titration bearing in mind the dichromate solution is 0·171N.

If the difference between the two ammonium ferrous sulphate titrations is x ml then

$$
\begin{aligned}
x \text{ ml of } y \text{ N-}(NH_4)_2Fe(SO_4)_2 &\equiv x \text{ ml } y \text{ N-}K_2Cr_2O_7 \\
&= 0{\cdot}00675xy \text{ g cellulose} \\
\text{Total weight of } \beta + \gamma \text{ cellulose} &= 100 \times 0{\cdot}00675xy \text{ g} \\
&= 0{\cdot}675xy \text{ g}
\end{aligned}
$$

Now transfer 100 ml of the $\beta + \gamma$ filtrate to a conical flask and titrate with 2N-sulphuric acid using methyl orange indicator (t ml). To the remaining 800 ml of the $\beta + \gamma$ filtrate add $(8t + 5)$ ml of 2N-sulphuric acid and dilute to 1 litre. Allow the β-cellulose to settle out overnight.

By means of a pipette, transfer 100 ml of the solution of γ-cellulose taking care to avoid picking up any precipitate, to a 500 ml conical flask and oxidise with dichromate and titrate exactly as for the $\beta + \gamma$ determination.

$$
\text{Total weight of } \gamma \text{-cellulose} = 125 \times 0{\cdot}00675xy \text{ g}
$$

Express the α-, β- and γ-cellulose contents as percentages of the original cellulose (dry weight).

37　Absorption of methylene blue by cellulose

APPARATUS

Spekker photo-electric absorptiometer, mechanical shaker, No. 2 sintered glass funnel.

MATERIALS

Methylene Blue, 5,5-diethylbarbituric acid (Poison, *see* Note 1), various samples of cellulose (*see* Note 2).

THEORY

Pure cellulose has no affinity for the basic dye methylene blue.

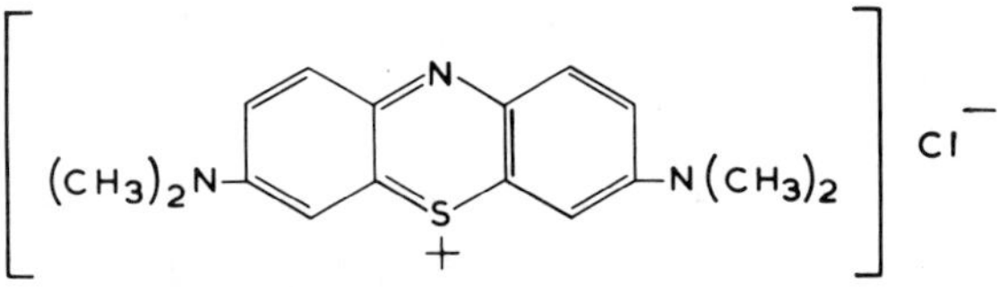

methylene blue (MB$^+$Cl$^-$)

Oxidation of cellulose to oxycelluloses however, may, but not invariably, produce materials containing carboxyl groups which absorb methylene blue (Peters [1]).

$$-COOH + MB^+Cl^- \rightleftharpoons -COO^-MB^+ + H^+ + Cl^-$$

The absorption of methylene blue is an ion-exchange mechanism, increasing with increasing methylene blue concentration and decreasing with increasing hydrogen ion concentration in accordance with the given equilibrium. To render the exchange nearly complete, it is necessary to use a solution of methylene blue buffered at a sufficiently high pH, but other cations such as Na^+ in the buffer solution will compete for the carboxyl groups. Nevertheless, it is possible by a suitable choice of experimental conditions to obtain a close approximation to equivalence between methylene blue absorption and the carboxyl content. It is preferable to exhaust about 50 per cent of the dye from a solution of pH 8 with a ratio of Na^+ to MB^+ of about $4:1$ at equilibrium (Davidson [2]).

EXPERIMENTAL PROCEDURE (*see* Note 3)

(*a*) *Preparation of methylene blue solution*—Prepare approximately 0·10N-sodium hydroxide by making up 4·0–4·1 g of the solid to 1 litre with distilled water. In a 1 litre graduated flask place methylene blue (1·280 g), diethylbarbituric acid (2·30 g) and 0·10N-sodium hydroxide (80 ml). Accurate weighing of the methylene blue is essential. Add distilled water, dissolve the contents, make up to the mark with distilled water and shake to ensure uniform mixing. Measure out accurately 100 ml of this solution and dilute to 1 litre with distilled water in a volumetric flask to give an exactly 0·40 mM solution with respect to methylene blue. This is the concentration used for the test.

(*b*) *Calibration of Spekker photo-electric absorptiometer*—Measure accurately 10, 25, 50 and 75 ml portions of the 0·40 mM-methylene blue solution and dilute each to 100 ml with distilled water in volumetric flasks. These solutions are of concentration 0·04, 0·10, 0·20 and 0·30 mM respectively. Dilute 10·0 ml of each solution, including the undiluted 0·40 mM solution, to 100 ml in a volumetric flask with approximately 0·1N-hydrochloric acid (8·9 ml conc. acid made up to 1 litre) and determine the optical density in the Spekker absorptiometer using a 1 cm glass cell and No. 8 red filter (*see* page 48). Plot a calibration graph of the Spekker reading against the original concentration (0·04–0·40 mM).

(*c*) *Absorption of methylene blue by cellulose*—Weigh accurately 0·1–2·0 g of cellulose, depending on the expected absorption. The latter will depend on the history of the sample or it may be necessary to perform a trial experiment (*see* later). The cellulose may be weighed dry after standing over phosphorus pentoxide, or the moisture content of another sample is determined by drying in an oven at 110°C and the dry weight of the material taken for the experiment is calculated from this. It is simpler, although not quite so accurate, to assume a moisture content of 6 per cent and determine the dry weight from this (*see* page xiii). Cut the sample into small pieces and shake for 2 h in a stoppered 100 ml flask with 0·40 mM methylene blue solution (100 ml).

Filter through a No. 2 sintered glass funnel. Take exactly 10 ml of the filtrate and dilute to 100 ml with 0·1N-hydrochloric acid in a volumetric flask. Measure the colour intensity (optical density) using the Spekker absorptiometer and obtain the final concentration of methylene blue from the calibration chart. Check the accuracy of the calibration by re-determining the concentration of the original 0·40 mM-methylene blue solution.

From Figure 20 it follows that the methylene blue absorbed is

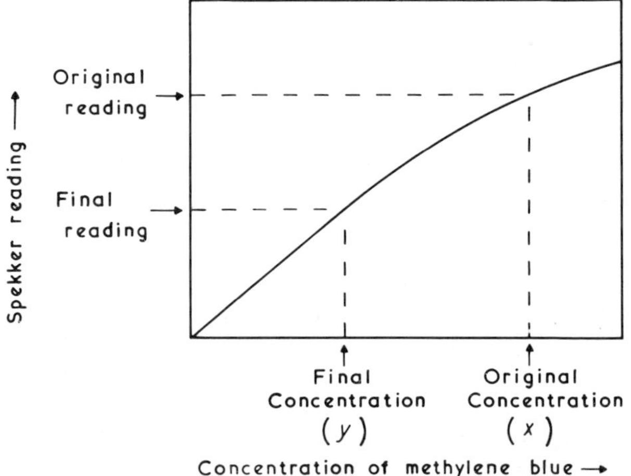

Figure 20.   Absorption of methylene blue from solution by cellulose

$(x - y)$ mg-mole per litre of solution. At equilibrium it is preferable that about 50 per cent of the methylene blue is exhausted and if y does not lie approximately midway between x and the origin, the determination should be repeated with a suitably adjusted weight of cellulose.

The result is calculated as follows, using w g of cellulose and 100 ml of methylene blue solution.

Methylene blue absorbed $= (x - y)$ mg-mol from 1 l of solution

$$= \frac{(x - y)}{10} \text{ mg-mol from 100 ml of solution}$$

Methylene blue absorbed by 100 g of cellulose $= \dfrac{10(x - y)}{w}$ mg-moles

Carboxyl content $= \dfrac{10(x - y)}{w}$ mg-moles per 100 g

The absorption of methylene blue may be assessed visually by washing the dyed cellulose, collected in the sintered-glass funnel, with distilled or de-ionised water until the filtrate is no longer coloured (tap-water cannot be used as calcium ions exchange with the MB^+ ions on the cellulose). The cellulose is air-dried and the intensities of colour retained by the different samples are compared.

NOTES

1. 5,5-Diethylbarbituric acid (Scheduled Poison) is known also as barbital, barbitone or veronal.

2. Suitable samples of oxy- and hydro-celluloses may be made as follows:

(i) Treat cotton linters with 10 per cent of their weight of available chlorine as sodium hypochlorite (*see* page 3) in pH 5 buffer solution ($0.05\text{M-NaH}_2\text{PO}_4$ + dropwise addition of N-NaOH) with a liquor ratio of 50:1 for 1 h at room temperature in a stoppered flask with occasional shaking. Filter on a Buchner funnel, wash well with water and air-dry.

(ii) Treat as in (i) for 24 h at room temperature replacing the phosphate buffer solution by 0.05M-borax of pH 9.2.

(iii) Reflux the linters for 1 h with $0.10\text{N-H}_2\text{SO}_4$ using a liquor ratio of 100:1. Filter and wash with water, dilute ammonia solution, water and air dry.

3. The abbreviation mg-mol stands for the weight of a substance equal to its molecular weight expressed in milligrammes. A solution containing this weight in a litre is of milli-molar concentration, which is abbreviated to mM.

REFERENCES

1 PETERS, R. H., *Textile Chemistry*, vol. I. Elsevier, Amsterdam, 209 (1962)
2 DAVIDSON, G. F., *J. Text. Inst.*, **39**, 65T (1948)

38 Fluidity of cellulose in cuprammonium hydroxide solution

APPARATUS

Fluidity tubes,* mechanical shaker, 0·25 in diameter steel ball bearings, 2 oz screw-capped polythene bottles, polythene sheet (from a bag is satisfactory).

MATERIALS

'Shirley' cuprammonium hydroxide solution (obtainable from laboratory suppliers), pyrogallol, samples of cotton (*see* Note 1) and viscose rayon.

DISCUSSION (*see* Onyon [1]).

All fluids possess resistance to change of form. This property, which may be regarded as being due to internal friction, is known as viscosity. Solutions of long chain molecules show an enhanced viscosity compared with the solvent itself, depending, among other factors, upon the chain length of the polymer. Thus, if a polymer has been degraded, its solution will show a reduction in viscosity compared with a solution of the intact polymer.

A convenient measure of the viscosity of a liquid is its time of flow through a capillary, and by the use of a suitable constant

* Also known as X-type viscometers, obtained from Shirley Developments Ltd., P.O. Box 6, 856 Wilmslow Road, Manchester 20.

applicable to the viscometer used in the determination (the visco-meter constant) this time may be converted into absolute units of viscosity (poises). The fluidity (reciprocal of viscosity) of a solution of cellulose in a suitable solvent is commonly used as a measure of the degradation it has suffered during chemical treatments, such as bleaching, which may shorten the molecular chains.

In view of the fact that the established procedure (*see* Clibbens and Geake [2] and Clibbens and Little [3]) involves long agitation of the cellulose with the solvent, the method described is a modi-fication (*see* Shirley Inst. Bull. [4]) which gives only slightly less accurate results in a much shorter time.

EXPERIMENTAL PROCEDURE (*See* Note 2)

(*a*) *Fluidity of cotton* —Yarn should be in singles form and cloth dissected into warp and weft yarns. The material should either be dried in an oven at 110°C for 1 h and cooled in a desiccator or it should be allowed to stand in a desiccator for 2 days over phos-phorus pentoxide. Then 0·125 g of dry material should be used or the cotton may be weighed under laboratory conditions and the assumption made that it contains 6 per cent moisture. In the latter case, 0·133 g will be required. Place the cotton in a 2 oz polythene bottle, add twelve 0·25 in diameter steel ball bearings, cupram-monium hydroxide solution (50·0 ml) and six drops of pyrogallol solution (10 g solid dissolved in 20 ml of water). Screw on the cap, with a layer of polythene sheeting between the bottle and cap to make a more effective seal. Shake vigorously for 30 min.

Transfer the 0·25 per cent solution of cotton to one or two fluidity tubes and seal the top only, using a clip. Hang the tube vertically in its glass jacket (*see* Figure 21) in a water bath at 20 ± 2°C (a 500 ml measuring cylinder is satisfactory) and run out the solution noting the time of outflow. Refer to the conversion chart (Figure 22) to find the fluidity of a 0·5 per cent solution in cuprammonium hydroxide solution which is the standard concentration for ex-pressing fluidities.

(*b*) *Fluidity of viscose*—The procedure is the same as for cotton except that 0·500 g of dry viscose or 0·556 g in equilibrium with atmospheric conditions is dissolved in 50·0 ml of cuprammonium hydroxide to give a 1·00 per cent solution. By means of Figure 21 the fluidity of a solution of double this concentration is obtained.

(*c*) *Calculation and significance of results*—If *t* is the observed time of flow in seconds between the top (B) and bottom (D) timing marks, the fluidity F of the solution in poises is given by

$$F = \frac{c}{t}$$

where c is the viscometer constant
or, if it is necessary to apply the kinetic energy correction, k,

$$F = \frac{c}{t - (k/t)}$$

The kinetic energy correction is unnecessary for times of flow greater than 200 s.

Chemically undamaged cotton has a fluidity of about 3 to 5.

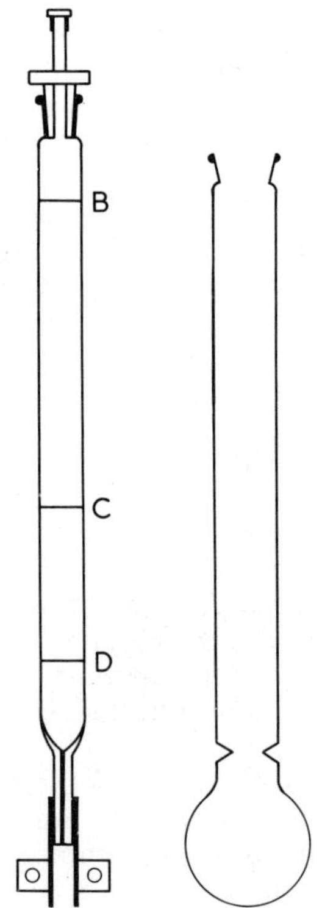

Figure 21. Shirley X-type viscometer and jacket

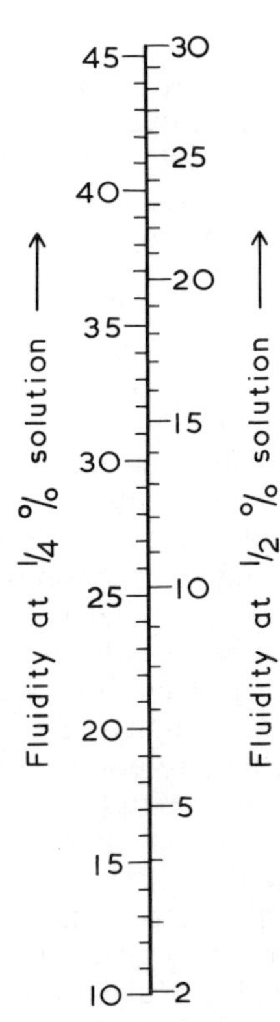

Figure 22. Fluidity conversion chart

A value of 10 or more can be taken as an indication of excessive degradation. Since the cellulose used in making viscose has suffered degradation during solution and regeneration, it is usual to work with a 2 per cent solution to bring the fluidities into line with those of cotton. Unprocessed viscose normally has a fluidity of 10 to 12.

NOTES

1. It is unsatisfactory to measure the fluidity of raw cotton since it contains at least 7 per cent of non-cellulosic material. All cotton tested must be well scoured. Cotton samples may be bleached in 1–2 vol. hydrogen peroxide solution brought to a pH of 10–11 with sodium silicate for 2 h at 85°C or in sodium hypochlorite (2–3 g available chlorine per litre) maintained at pH 10–11 with sodium carbonate (5 g/l) for 2 h at 20°C.

2. In the standard procedure, as distinct from the rapid method, the cellulose solution is prepared in the viscometer tube. Each tube is therefore individually calibrated to give the exact weight of cellulose necessary to yield a solution of the required concentration when the tube is completely filled. Since in the determination described, the solution is prepared separately, this weight of cellulose is ignored. The three marks on the tube are timing marks. Observations at the middle mark C are not used in calculating the result and this mark may be used to detect large errors due to possible obstruction of the capillary. If t_1 is the time of flow from the top (B) to the middle mark (C) and t_2 the time of flow from the middle (C) to the bottom mark (D), the ratio t_1/t_2 should be near unity. (Differences between t_1 and t_2 of less than 10 per cent may be disregarded as these are due to deviation from true viscous flow).

When viscosity measurements are made in order to determine the molecular weight of a polymer, temperature control is extremely critical. Sufficient accuracy however, is attained in these determinations if the temperature is within the range 18–22°C.

REFERENCES

1 ONYON, P. F., *Techniques of Polymer Characterization.* (Ed. by P. W. Allen) Butterworths, London, 171 (1959)
2 CLIBBENS, D. A. and GEAKE, A., *J. Text. Inst.,* **19**, T77 (1928)
3 CLIBBENS, D. A. and LITTLE, A. H., *J. Text. Inst.,* **27**, T285 (1936)
4 *Shirley Institute Bulletin.* Manchester, England, **39**, 29 (1966)

39 Determination of the copper number of cellulose

APPARATUS

No. 2 porosity sintered glass filter crucible.

MATERIALS

Anhydrous sodium carbonate, sodium bicarbonate, copper sulphate ($CuSO_4 \cdot 5H_2O$), ferric alum (ammonium ferric sulphate), ammonium ceric sulphate, ammonium ferrous sulphate, potassium dichromate, 90 per cent orthophosphoric acid. Ferroin indicator, sodium diphenylamine sulphonate indicator (*see* Note 1). Samples of cellulose, for example wood pulp, bleached linters, hypochlorite-treated cotton (*see* page 129).

THEORY

Hydrolytic breakdown of cellulose across the oxygen bridges of the glucose residues results in the formation of two hydroxyl groups. Whereas the hydroxyl group attached to the 4-carbon atom is non-reducing, that attached to the 1-carbon atom is a potential aldehyde group (*see* Peters [1]). Oxidation of cellulose can produce ring fission of the glucose residues, resulting in the formation of aldehyde groups at carbon atoms 2 and 3 (*see* Peters [1] and page 112 for the method of numbering the carbon atoms in the glucose ring). These three aldehyde groups can reduce Fehling's solution to cuprous oxide. The latter may be determined quantitatively by allowing the oxide to reduce ferric alum to the ferrous state and determining the latter by titration with ammonium ceric sulphate. The copper number is the weight of copper reduced from the Cu^{2+} to the Cu^+ state by

100 g of dry cellulose and is a measure of its inter and intra chain breakdown.

EXPERIMENTAL PROCEDURE

(*a*) *Preparation of solutions*—Six solutions are required for this determination and these are made as follows:

(i) *Sodium carbonate*—Add anhydrous sodium carbonate (130 g) to distilled water (750 ml) at about 70°C and stir to effect solution. Dissolve sodium bicarbonate (50 g) in the solution and then filter through a Buchner funnel and filter paper and make up the cool solution to 1 litre.

(ii) *Copper sulphate*—Dissolve hydrated cupric sulphate (25 g) in distilled water and make up to 250 ml.

(iii) *Ammonium ferric sulphate (Ferric alum)*—Dissolve ferric alum (50 g) in distilled water (250 ml). Add conc. sulphuric acid (70 ml) slowly with stirring and cooling. Dilute with water to 500 ml.

(iv) *0·05 N-Potassium dichromate* (*see* Note 2)—Potassium dichromate oxidises by the following mechanism

$$K_2Cr_2O_7 + 4H_2SO_4 \rightarrow K_2SO_4 + Cr_2(SO_4)_3 + 4H_2O + 3O$$

$$\therefore \text{Equivalent weight} = \frac{\text{Molecular weight}}{6} = 49\cdot03$$

Potassium dichromate is a primary standard and a standard solution can be prepared by direct weighing. Weigh accurately 2·4–2·6 g dry A.R. potassium dichromate in a weighing bottle. Transfer to a 1 litre volumetric flask by means of a funnel and distilled water. Dissolve, make up to the mark with distilled water and shake well. Calculate the exact normality.

(v) *0·05 N Ammonium ferrous sulphate* (*see* Note 2)—Dissolve ammonium ferrous sulphate (5·0 g) in water (approximately 150 ml) add 2N-sulphuric acid solution (25 ml) and make up to 250 ml with distilled water. Shake to ensure uniform mixing.

To standardise this solution, pipette 25·0 ml into a 500 ml conical flask and add sodium diphenylamine sulphonate indicator (8 drops), water (100 ml), 2N-sulphuric acid solution (100 ml) and 90 per cent orthophosphoric acid (5 ml). Titrate slowly with the standard dichromate solution with constant shaking until the solution assumes a blue-green tint. Continue adding dichromate solution dropwise at intervals of a few seconds until one drop causes the formation of an intense purple or violet colour. Calculate the normality of the ammonium ferrous sulphate solution.

(vi) *Ammonium ceric sulphate (approximately 0·04N)*—In a tall 250 ml beaker mix together ammonium ceric sulphate (7·5 g) and concentrated sulphuric acid (13 ml) and heat gently while stirring until the acid just begins to fume. The product should be a yellow paste free from lumps. *Allow to cool* and dilute by the cautious addition of water (125 ml). Dilute to 250 ml and shake well. This solution is standardised with solution (v). Transfer 25·0 ml of ammonium ferrous sulphate solution to a 250 ml conical flask. Add 2N-sulphuric acid solution (25 ml) and ferrous-*o*-phenanthroline (Ferroin) indicator (2–3 drops). Titrate with the ammonium ceric sulphate solution to the first sharp colour change (orange-red to pale blue) and calculate its normality.

(*b*) *Determination of copper number*—Weigh accurately about 2·5 g of the cellulosic material and place in a 100 ml stoppered conical flask. The material which has previously been well divided and cut into small pieces, is weighed in equilibrium with normal atmospheric conditions (*see* Note 3). In a 250 ml beaker, mix carbonate solution (i) (95 ml) and copper solution (ii) (5 ml) and raise to the boil. Pour immediately into the flask containing the cellulose and mix thoroughly. Stopper, and immerse the flask to the neck in a bath of boiling water for 3 h. Agitate occasionally during the first 15 min to release air bubbles.

Filter the contents through a No. 2 porosity sintered glass crucible. Wash out the flask with some carbonate solution (i) and pour over the cellulose in the filter. Repeat with hot water (100 ml), finally pressing the sample with a flattened glass rod. Nearly fill the crucible six times with further quantities of hot water, allowing it to drain thoroughly and pressing the cellulose between each addition. Reject the filtrate and washings.

Fit the crucible into the neck of a clean 250 ml filter flask and dissolve out the cuprous oxide by treatment with three successive portions (each 10–15 ml) of ferric alum solution (iii), using the first portion to remove any cuprous oxide from the original 100 ml flask before pouring into the crucible. Wash the material finally with two portions of 2N-sulphuric acid solution (each 15 ml). During each of the latter five washes, release the suction to allow the solution to permeate the cellulose.

Titrate the whole of the filtrate with the standardised ammonium ceric sulphate solution (vi), using ferroin indicator (2–3 drops), until the colour changes from red to pale blue.

The cuprous oxide reduces ferric alum as follows:

$$Cu^+ + Fe^{3+} \rightarrow Cu^{2+} + Fe^{2+}$$

and the final titration with ammonium ceric sulphate involves the oxidation $Fe^{2+} \rightarrow Fe^{3+}$

$\therefore$ 1 litre of N ammonium ceric sulphate solution $\equiv$ 63·54 g copper reduced.

or 1·0 ml of N ammonium ceric sulphate solution $\equiv$ 0·06354 g copper reduced.

$$\therefore \quad \text{Copper number} = \frac{V \times N \times 0.06354 \times 100}{W \times C}$$

where V = volume of ammonium ceric sulphate solution of normality N

W = weight of cellulose

C = correction factor for moisture content of cellulose (0·94) (*see* Note 3).

NOTES

1. Sodium diphenylamine sulphonate indicator is made by dissolving the solid (0·2 g) in distilled water (100 ml). Ferroin indicator may be purchased or it can be prepared by dissolving ortho-phenanthroline monohydrate (1·485 g) and ferrous sulphate ($FeSO_4 \cdot 7H_2O$) (0·695 g) in distilled water (100 ml), without warming.

2. A.R. Ammonium ferrous sulphate is at least 99 per cent pure and no great error will be introduced into the determination if it is used as a primary standard. In this case, weigh the ammonium ferrous sulphate accurately (approximately 5·0 g), make up to 250 ml with water and calculate the exact normality of the solution (equivalent weight = 392·1) Alternatively, an exactly 0·050N solution may be prepared by making up 4·901 g of ammonium ferrous sulphate to 250 ml with water. It is unnecessary to prepare potassium dichromate solution (iv) if this procedure is adopted.

3. To avoid this factor, the cellulose may be dried under vacuum over phosphorus pentoxide and weighed in a stoppered bottle.

REFERENCE
1 PETERS, R. H., *Textile Chemistry,* vol. I. Elsvier, Amsterdam, 199, 211 (1963)

40 Mercerising of cotton

APPARATUS

No special apparatus is necessary for experiment (a). For experiment (b), a yarn (single thread) strength tester and a microscope ($\times 100$) are required. As a safety measure, hard glass test-tubes and boiling-tubes should be used.

MATERIALS

Acetic acid, iodine, potassium iodide, Chlorazol Blue or other suitable direct cotton dye, cotton yarn and fabric.

THEORY

Native cellulose (Cellulose I) forms alkali cellulose I with conc. sodium hydroxide solution. On washing and neutralisation, cellulose II is formed.

$$C_6H_7O_2(OH)_3 + NaOH \rightarrow C_6H_7O_2(OH)_2(ONa)$$

Cellulose I   Alkali cellulose I

$$\downarrow H_2O$$

$$C_6H_7O_2(OH)_3 + NaOH$$

Cellulose II

As a result of the penetration of alkali into the lattice, internal hydrogen bonds are broken and in Cellulose II the number of available hydroxyl groups is increased by 25 per cent. The process is known commercially as mercerisation. The treatment with alkali and subsequent washing may be performed so that the fabric or yarn may either freely contract or they may be held under tension. In both cases the mercerised cotton has an increased affinity for direct cotton dyes and water and an increased strength. Cotton yarn or fabric mercerised without tension contracts, but if held under tension it retains its original dimensions and the lustre is increased.

EXPERIMENTAL PROCEDURE

(*a*) *Mercerisation of fabric without tension*—Weigh four patterns of cotton fabric (8 cm × 8 cm approximately) to the nearest centi-gramme (combined weight). Coat the edges with cellulose acetate–cadmium sulphide marking fluid (*see* page 62) to prevent fraying. Wet out in water containing a few drops of 'Teepol' for 5 min and then squeeze well to remove as much water as possible. Measure the mean length and breadth of two samples to obtain their initial areas.

Immerse the two measured patterns in sodium hydroxide solution (30 g sodium hydroxide + 100 ml water) for 15 min at room temperature with stirring (CAUTION—this solution is extremely caustic), remove the samples (**retain** the caustic soda solution) and wash thoroughly with water and then stir with 1 per cent acetic acid solution (400 ml) at 40°C for 5 min to neutralise the absorbed alkali. Rinse well and measure. Calculate for each sample the percentage reduction in area.

Dye one untreated and one alkali treated sample together in a bath containing: Water (400 ml), Sodium chloride (10 per cent on weight of fabric), Chlorazol Blue (1 per cent on weight of fabric). Raise to 60°C before introducing the fabric, bring to the boil and boil for 30 min. Rinse, dry and mount the samples for comparison.

Take the two remaining patterns, dry for 1 h at 110°C and weigh accurately on an analytical balance after allowing to cool in a desiccator. Allow to stand for 48 h either in the laboratory atmosphere or preferably in a desiccator over saturated sodium nitrite solution. The atmosphere over the latter will have a relative humidity of 66 per cent at 20°C. Determine the percentage water absorption on the untreated and mercerised cottons.

Saturate water (10 ml) with potassium iodide and shake with iodine (2 g) to effect solution. Take strips of untreated and mercerised fabric (from the water absorption determination) wet out, wash off and squeeze. Immerse these for 3 s in the iodine-potassium iodide solution and wash off very thoroughly with cold water. Record the coloration of each sample.

(*b*) *Mercerisation of yarn under tension*—Take a 5 in × $\frac{5}{8}$ in hard glass test-tube and cover the exterior with an evenly wound layer of cotton yarn to within an inch of the open end. Wet out the yarn on the tube as previously and remove excess water with filter paper. Place about 30 ml of mercerising solution which may have been retained from experiment (*a*), in a 6 in × 1 in hard glass boiling-tube and insert a test-tube brush into the small test-tube. Using this as a handle, insert the small test-tube and yarn into the boiling-tube and treat for 15 min at room temperature with occasional agitation. Wash and neutralise the yarn as in (*a*) and allow to air-dry on the tube.

The sodium hydroxide solution should be discarded after this experiment.

The yarn is removed from the tube and its appearance compared with that of untreated yarn. A few fibres from each should be examined under the microscope (× 100) and sketches made of un-mercerised and mercerised cotton fibres.

The breaking load and extensibility of mercerised yarn should be compared with those of untreated cotton by means of a yarn strength testing machine and the mean of as many results as possible should be taken. Normally the difference in strength is so marked that statistical analysis of the results is unnecessary.

REFERENCE
1 MARSH, J. T., *Mercerizing*. Chapman & Hall, London (1949)

41 Preparation of cellulose triacetate and its hydrolysis to secondary acetates

APPARATUS

No special apparatus is required.

MATERIALS

Acetylation-grade linters,* glacial acetic acid, acetic anhydride, 90 per cent formic acid, acetone, chloroform, 1.0N-sodium hydroxide of exactly known normality, 1.0N-hydrochloric acid (approximate normality only required).

THEORY

The acetates of cellulose are used widely in the plastics and man-made fibre industries.

The triacetate is prepared by the complete acetylation of cellulose, generally in the presence of sulphuric acid as catalyst.

$$C_6H_7O_2(OH)_3 + 3 \begin{matrix} CH_3CO \\ \diagdown \\ \diagup \\ CH_3CO \end{matrix} O \rightarrow$$

$$\rightarrow C_6H_7O_2(OCOCH_3)_3 + 3CH_3COOH$$

Cellulose triacetate

* Suitable material is supplied by Alfred A. Brown Ltd., Roman Ridge Road, Wincobank, Sheffield, England.

Although this may be used as such, for example 'Tricel' yarn, often the diacetate is required. This cannot be obtained satisfactorily by the partial acetylation of cellulose, but only by partial hydrolysis of the fully esterified material.

$$C_6H_7O_2(OCOCH_3)_3 + H_2O \rightarrow$$

Cellulose triacetate

$$\rightarrow C_6H_7O_2(OH)(OCOCH_3)_2 + CH_3COOH$$

Cellulose diacetat

The latter forms the basis of ordinary cellulose acetate yarn ('Dicel').

EXPERIMENTAL PROCEDURE

(*a*) *Pre-treatment of cellulose*—Tease out the linters with fine-pointed tweezers and dry for 1 h at 110°C. prior to weighing. Place the linters (20·0 g) in a 1 litre beaker and add 90 per cent formic acid (4·0 ml) dropwise from a burette over 15 min with constant agitation of the linters. Cover with a well-fitting clock-glass and allow to stand 2 h.

(*b*) *Acetylation of cellulose*—Prepare a mixture of acetic anhydride (60 ml), acetic acid (140 ml) and conc. sulphuric acid (1·6 ml) in a 500 ml conical flask. Cool to 5°C in water containing ice (sometimes crystallisation commences at this temperature but this is not important). Place the pre-treated cellulose in a 600 ml beaker, add the acetylation mixture and knead with a thick glass rod pressed into a disk at the end. After 20–25 min a maximum temperature should be reached of 40–45°C. It is essential to reach this temperature but if it is exceeded, cool to this temperature immediately. Continue to agitate for 5 min and then allow to stand for 2 h.

A completely clear solution devoid of any unreacted cellulose must be obtained at this stage. If this is not so, the acetylation must be repeated.

(*c*) *Hydrolysis of the triacetate* (*ripening*)—To the solution of the triacetate add 60 per cent (v/v) aqueous acetic acid (30 ml). Stir thoroughly for 10 min and note the rise in temperature from the reaction between the excess anhydride and water.

$$
\begin{array}{c}
CH_3CO \\
\diagdown \\
\diagup \\
CH_3CO
\end{array}
O + H_2O \rightarrow 2CH_3COOH
$$

Allow to stand 2 h and regard this as zero time of hydrolysis. If possible, allow to stand in a controlled temperature of 20°C. Precipitate approximately 40 ml of solution after 0, 24, 48, 72 and 96 h, or at other suitable times to fit in with the laboratory periods available.

(*d*) *Precipitation of cellulose acetates*—The object of this operation is to precipitate a fibrous material which is devoid of both powder which may clog the filter and lumps which may contain gels. As ripening proceeds, it will probably be found that less fine material is precipitated and the acetate becomes easier to filter and wash.

Pour the acetate solution (approx. 40–50 ml) into a 800 ml beaker and beat in very vigorously water (40–80 ml) in 10 ml portions to bring to precipitation point. Then add 600 ml of water quickly stirring slowly. Precipitate the final 96 h sample in the original beaker.

Filter through paper in a Buchner funnel under vacuum if necessary, and when drained completely, wash with cold water (1 litre), followed by hot water (70°C) until three consecutive washings are neutral to indicator paper. Wash with a further 2 litres of hot water and allow to drain thoroughly. Air-dry on filter paper for 24 h and then in a vacuum desiccator over silica gel. The unstabilised acetate prepared, contains sulphate groups and must not be dried at high temperature.

Determine the total yield from all the precipitations and express as a percentage of the theory (*see* Note 1).

On each sample of acetate observe the effect of acetone and chloroform (*see* Note 2) and determine the acetyl value.

(*e*) *Determination of acetyl value* (*see* Note 3)—This is defined as the percentage weight of acid liberated when the acetate is completely hydrolysed to cellulose and acetic acid. A different acetyl value is defined as the percentage weight of acetyl group, $CH_3CO\cdot$, in the ester, but the former definition is nearly always used in industry. Acetyl values, which should be performed in duplicate, are determined as follows:

Place approximately 2 g of the thoroughly dry acetate in an unstoppered weighing bottle and leave in a vacuum desiccator over phosphorus pentoxide for at least 24 h. Weigh and transfer the contents to a 500 ml flask and reweigh the bottle. Add 150 ml aqueous acetone and warm on a water bath if necessary to complete solution (no naked flames). For zero time of hydrolysis use 95/5 V/V acetone–water, for 24, 48 and 72 h. Use 90/10 V/V acetone–water and for 96 h use 85/15 V/V acetone–water. As hydrolysis proceeds, the acetate will become more readily soluble and heat will not be required to effect solution.

For the titration, N-sodium hydroxide of exactly known normality is required but the N-hydrochloric acid need not be standardised. 1 litre of each should be prepared (*see* Note 4). Add exactly 30 ml N-sodium hydroxide and 20 ml water for hydrolysis. Add slowly with shaking to obtain a fine dispersed precipitate of cellulose. To a blank of 150 ml of the same aq. acetone add exactly the same volume of N-sodium hydroxide and water. Titrate this blank with N-hydrochloric acid using phenolphthalein. After standing for a few minutes to complete the hydrolysis, add exactly the same volume of N-hydrochloric acid to the determination. Titrate the remaining acetic acid with N-sodium hydroxide using phenolphthalein indicator.

$$\text{A.V.} = \frac{6 \times \text{volume } 1\cdot0\text{N-NaOH soln. (ml)}}{\text{weight of sample (g)}}$$

NOTES

1. The acetates produced will on average correspond to $C_6H_7O_2(OH)_{0\cdot5}(OCOCH_3)_{2\cdot5}$. Therefore, 20 g of cellulose will give a theoretical yield of 33 g. A yield of at least 90 per cent of this figure should be obtained.

2. As hydrolysis proceeds, the solubility of the acetate in chloroform decreases and that in acetone increases. Over a small range of acetyl values, this effect may not be too apparent.

3. Duplicate acetyl values should have a maximum range of $\pm 0\cdot2$ per cent. The theoretical value for cellulose triacetate is $62\cdot5$ per cent and over 96 h this should drop to about 54 per cent.

4. Exactly N-sodium hydroxide may conveniently be prepared from ampoules, for example 'Volucon' solutions supplied by May and Baker. Approximately N-hydrochloric acid is made by diluting 84 ml of the conc. acid to 1 litre with water.

REFERENCE
1 MONCRIEFF, R. W., *Man-made Fibres*. Heywood, London, 204 and 224 (1963)

42 Preparation and analysis of cellulose acetate-stearate

APPARATUS

Soxhlet extractor.

MATERIALS

Cellulose acetate (acetyl value approximately 54 per cent as acetic acid supplied by Hopkin and Williams or May and Baker is suitable), stearic anhydride, acetic anhydride, ethylene dichloride, acetone, Industrial Methylated Spirits, petroleum ether (b.p. 40–60°C), anhydrous (fused) zinc chloride, freshly ignited anhydrous magnesium sulphate, 90 per cent orthophosphoric acid, volumetric solutions for acetyl values (*see* page 141).

CAUTION Acetone and petroleum ether are highly inflammable and ethylene dichloride vapour is both inflammable and toxic.

DISCUSSION

Although cellulose acetate is the most important organic ester of cellulose, being used widely in the manufacture of fibres and plastics, other simple and mixed esters are also used for such purposes as moulding powders and lacquers, although they have not achieved the importance of the acetates.

In the preparation described, cellulose acetate is reacted with stearic anhydride in the presence of zinc chloride as catalyst, to produce a mixed acetate–stearate by ester interchange.

EXPERIMENTAL PROCEDURE

(*a*) *Stearylation of cellulose acetate*—The solvents used in this preparation should be dried by adding anhydrous magnesium sulphate and standing for several days with occasional agitation. They are separated from the solid by filtration.

To prepare the esterification catalyst, anhydrous zinc chloride (2·2 g) is dissolved in acetone (25 ml). Concentrated hydrochloric acid (2·0 ml) is added followed by ethylene dichloride (25 ml). The solution is dried over anhydrous magnesium sulphate and must be used within a day of its preparation.

The stearylation mixture consists of cellulose acetate dried for at least 1 h at 110°C (23·0 g), stearic anhydride (30 g), ethylene dichloride (150 ml) acetic anhydride (2·0 ml) and freshly prepared catalyst (5·0 ml). The semi-solid mass, which is contained in a 500 ml conical flask with glass stopper is broken up and mixed using a thick glass rod. The closed flask is placed in an oven at 45–50°C for 4 h, being stirred after the first hour and shaken each subsequent hour. Lumps are broken up with a thick glass rod flattened into a disk at the end. Pour the viscous solution through a warm funnel containing a loose plug of glass wool. Warm the filtrate gently if it solidifies and then, with vigorous stirring, pour into a litre of boiling water contained in a 2 litre beaker (*see* Note). This precipitation and subsequent boiling must be performed in a fume cupboard. Continue boiling for 15 min when the viscous semi-solid should solidify completely.

Remove the solid by filtration through a Buchner funnel, break up into pieces about 1 cm cube and dry with filter paper. Air-dry for at least 24 h. Extract in a Soxhlet apparatus using a thimble for 4 h with 40–60°C petroleum ether on a water bath. Allow to air-dry until free from the odour of solvent and finally dry for 1 h at 110°C. Weigh and record the yield.

(*b*) *Analysis of cellulose-acetate stearate*—

1. Determine the acetyl value of the original cellulose acetate as described on page 143 dissolving in 85/15 acetone–water. Note that the acetyl value is the number of grammes of acetic acid liberated by 100 g of the ester.

2. Determine the acetyl value of the cellulose acetate–stearate

using the same procedure but dissolving in 100 per cent acetone. The stearate group does not titrate under these conditions.

3. Determine the stearyl value of a sample of cellulose acetate–stearate by the following method.

Reflux an accurately known weight (approximately 1 g) of the dry product with 90 per cent orthophosphoric acid (20 ml) for 1·5 h. Allow to cool and filter through a sintered glass crucible. Wash the residue of stearic acid with water (100 ml). Dissolve the stearic acid remaining in the condenser and in the crucible in hot alcohol (2 × 15 ml) and add this solution to water (200 ml). Filter off the precipitated stearic acid on to a sintered glass crucible and wash well with distilled water. Dissolve in alcohol (25 ml) and titrate with 0·10N-sodium hydroxide using phenolphthalein as indicator.

If w g of acetate–stearate require T ml of alkali,

$$\text{Stearyl value} = \frac{2 \cdot 84T}{w}$$

Comment on the results.

NOTE

In view of the fact that there is a fire risk with ethylene dichloride, the precipitation must be performed in the absence of naked flames. The subsequent boiling may be carried out over a small flame. If the vapour ignites, it may be extinguished by placing an asbestos centred gauze or a sheet of asbestos over the beaker.

L

43   End-group analysis of starch by periodate oxidation

APPARATUS

10 ml microburette.

MATERIALS

Potassium chloride, sodium metaperiodate, ethylene glycol, amylose and amylopectin,* methyl red indicator.

THEORY

In general, the long chain molecules of polysaccharides such as starch consist of glucose residues linked through their 1:4 positions. (*See* page 112 for the method for numbering the carbon atoms). When the chain is terminated by a glucose residue with the C–1 position (potential aldehyde) free, this group is reducing, but if the C–4 position is free it will be non-reducing. If starch is oxidised with periodate, one molecule of formic acid is produced from each non-reducing end-unit and two molecules are produced from each reducing end-unit.

Starch may be fractionated with hot water into amylose (soluble) and amylopectin (insoluble). The molecular weight by osmotic pressure of the former corresponds to its end-group assay and shows it to be an unbranched long-chain molecule. On the other hand,

* The inexpensive materials produced by Koch-Light Laboratories Ltd., Colnbrook, Buckinghamshire, England, give satisfactory results.

although the molecular weight of amylopectin indicates that it
must consist of a minimum of 300 glucose residues, end-group

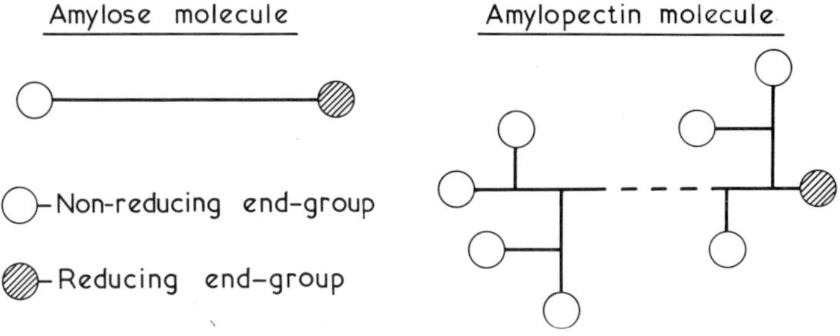

Figure 23.   Outline molecular structure of amylose and amylopectin

analysis shows the presence of one non-reducing end-group for
each 25 units, indicating that the molecule must consist of branched
chains (Figure 23). These branching points are across carbon atoms
1 and 6 (*see* Fieser and Fieser [1]).

From Figure 23 it follows that each amylose molecule will yield
three molecules of formic acid and thus the latter may be used to
determine its molecular weight. In the case of amylopectin, where
it may be assumed that all the formic acid arises from non-reducing
end-groups, the yield of this acid may be used to determine the
average molecular weight of the branches, but not the overall
molecular weight of the carbohydrate.

EXPERIMENTAL PROCEDURE (*see* Shasha and Whistler [2])

The following solutions are required:

(*a*) 5 Per cent potassium chloride. (Make up potassium chloride
(50 g) to 1 litre with water).
(*b*) 0·3M-sodium periodate (dissolve sodium metaperiodate (6·42 g)
in water, make up to 100 ml in a volumetric flask and store in
the dark).
(*c*) 0·010N-sodium hydroxide. Prepare 0·10N-sodium hydroxide
from the solid (*see* page 73) or from an ampoule and dilute 100 ml
to 1 litre with water in a volumetric flask.

Dry the amylose or amylopectin (approximately 0·5 g) over phosphorus pentoxide in an evacuated desiccator for 3 days and weigh.

Place in a 500 ml stoppered flask and add 5 per cent potassium chloride solution (exactly 90 ml) and 0·3M-sodium periodate (exactly 10 ml). Prepare a blank solution containing the same quantities of potassium chloride and sodium periodate solutions, but without the amylose or amylopectin. Keep the solutions in the dark at room temperature with occasional shaking for 14 days.

Withdraw 25 ml portions of the solutions, mix each with 1 ml of ethylene glycol and allow to stand for 10 min in the absence of light to remove excess periodate. Finally, titrate the formic acid present with 0·010N-sodium hydroxide using a 10 ml microburette and methyl red indicator.

Calculation of results—Let x g of carbohydrate require y ml of 0·010N-NaOH (corrected for blank) for neutralisation of the formic acid produced on oxidation.

Since,

$$1{\cdot}0 \text{ ml } 0{\cdot}010\text{N-NaOH} = 0{\cdot}010 \text{ mg-mol formic acid}$$

$$\text{Total formic acid produced by sample} = 0{\cdot}04y \text{ mg-mol}$$

(*a*) *Amylose*—

$$\text{Non-reducing end-groups (N)} = \frac{0{\cdot}04y \times 100}{3x} \text{ mg-mol/100 g}$$

$$= \frac{1{\cdot}333y}{x} \text{ mg-mol/100 g}$$

$$\text{Molecular weight } (M) = \frac{10^5}{N}$$

$$\text{No. of glucose units per molecule} = \frac{M}{162}$$

(*b*) *Amylopectin*—

$$\text{Non-reducing end-groups } (N) = \frac{0{\cdot}04y \times 100}{x} \text{ mg-mol/100 g}$$

$$= \frac{4y}{x} \text{ mg-mol/100 g}$$

$$\text{Average molecular weight per branch } (M) = \frac{10^5}{N}$$

$$\text{Average number of glucose units per branch} = \frac{M}{162}$$

REFERENCES

1 FIESER, L. F. and FIESER, M., *Organic Chemistry*. Reinhold, New York, 389 (1956)
2 SHASHA, B. and WHISTLER, R. L., *Methods in Carbohydrate Chemistry*. Vol. IV (*Starch*) Academic Press, New York, 86 (1964)

PART 4   Experiments on Synthetic Polymers

44 Preparation of urea-formaldehyde resin

APPARATUS

Apparatus necessary to perform a Kjeldahl nitrogen determination (*see* page 23) and distillation under reduced pressure (*see* page 89), 250 ml three-necked flask, mechanical stirrer.

MATERIALS

Forty per cent formaldehyde solution (Formalin), urea, acetic acid, 90 per cent orthophosphoric acid, sodium sulphite, 1·00N-sulphuric acid, phenolphthalein indicator, pH test papers, cotton fabric, materials for Kjeldahl nitrogen analysis (*see* page 23) and, optionally, melamine.

DISCUSSION

Urea-formaldehyde resins are applied widely to fabrics made of cotton and staple viscose. The finish, which was invented by Tootal Broadhurst Lee Co., gives the material a fuller handle and enhances its resistance to creasing.

Depending on the proportions used, urea and formaldehyde react to produce mono- or dimethylol urea.

$$H_2NCONH_2 + HCHO = H_2NCONHCH_2OH$$
Monomethylol urea

$$H_2NCONH_2 + 2HCHO = HOCH_2NHCONHCH_2OH$$
Dimethylol urea

On heating, these products condense with the elimination of

water to produce cross-linked structures which are infusible and insoluble in non-destructive solvents.

The reactions are complex but linkages of the following types are probably formed:

$$—CH_2OH + HOCH_2— \quad \rightarrow —CH_2OCH_2— \ + H_2O$$

$$—CH_2OH + H_2NCO— \quad \rightarrow —CH_2NHCO— + H_2O$$

$$—CH_2OH + H_2NCO— \rightarrow —CH_2\overset{\displaystyle CH_2}{\underset{|}{\overset{|}{N}}}CO— \ + H_2O$$

The urea-formaldehyde is applied to the fabric as soluble methylol derivatives or low molecular weight precondensates. The impregnated fabric is then baked (cured) to complete the condensation.

EXPERIMENTAL PROCEDURE

(a) *Preparation of urea-formaldehyde precondensate*—Dissolve sodium hydroxide (1 g) in water (10 ml) and add dropwise to the formaldehyde solution (60 ml) until the pH is 7·5 as shown by a test paper. Add this solution to urea (30 g) contained in a 250 ml three-necked flask fitted with a stirrer and reflux condenser. Gently reflux with stirring for 2 h. Concentrate the mixture to a solid content of 70 per cent by distilling off water (20 ml) under reduced pressure from a water pump using an air-leak to prevent bumping (*see* page 89). Prior to use, acidify the precondensate syrup to pH 5·0 with acetic acid.

(b) *Application to cotton fabric*—Weigh patterns of cotton fabric (approximately 10 cm × 10 cm) under normal atmospheric conditions and calculate their approximate dry weights (*see* page xiii). Take approximately 10 ml of the resin precondensate at pH 5·0, add 90 ml of water and stir well. Impregnate the patterns with this solution, remove the excess with rubber rollers or by hand squeezing and dry at 110°C. Finally, cure at 140°C for 5 min and determine their dry weight. Calculate the percentage weight of resin deposited on the cotton and note its effect on fabric properties such as handle and resistance to creasing.

(c) *Analysis of the resin*—Pour the remainder of the resin on to five 3 in watch glasses and heat in an oven for 3 h at 110°C. Remove the sheets of resin from the glass and determine the yield (corrected for that used under (a)).

Determine the nitrogen content by the Kjeldahl method (*see* page 23) and the formaldehyde content as follows.

Weigh accurately the dry resin (approximately 1 g) and add 10 per cent phosphoric acid solution made by diluting the conc. acid (13 ml) to 200 ml with water. Distil very slowly into a solution containing water (20 ml), sodium sulphite (5 g) and phenolphthalein indicator (a few drops), the sulphite solution being previously neutralised with N-sulphuric acid. Use the same apparatus (flask, splash-head, condenser and delivery tube) as for the Kjeldahl distillation and ensure that the tip of the delivery tube dips just below the surface of the sulphite solution. Distil 20–30 ml and continue with fresh portions of sulphite solution until no more formaldehyde is evolved as indicated by the lack of formation of alkali.

$$CH_2O + Na_2SO_3 + H_2O \rightarrow H_2C(OH)SO_3Na + NaOH$$

Titrate the total alkali formed with 1·00N-sulphuric acid and calculate the formaldehyde content of the resin as percentage CH_2O.

$$1\cdot0 \text{ ml } 1\cdot00\text{N-}H_2SO_4 \equiv 0\cdot030 \text{ g } CH_2O$$

Calculate the percentages of nitrogen and formaldehyde which would be present in the following:

$$-NHCONHCH_2- \qquad\qquad -\overset{\displaystyle |}{N}CONCH_2-$$

$-NHCONHCH_2-$ $-NCONCH_2-$
       CH_2

Linear resin   Fully cross-linked resin

Relate the analytical figures obtained ·experimentally to these structures.

NOTE

A melamine-formaldehyde resin may be made by the procedure given using 40 per cent formaldehyde solution (20 ml) and melamine (20 g). It is unnecessary to concentrate by distillation. It is more stable than the urea-formaldehyde resin and to determine its formaldehyde content it is hydrolysed with 200 ml of 5N-sulphuric acid (140 ml conc. acid per litre) and after each distillation step, the water lost from the flask should be replaced.

REFERENCE
1 MARSH, J. T., *Self-Smoothing Fabrics.* Chapman & Hall, London, 45–63 (1962)

45 Preparation of polystyrene

APPARATUS

Ostwald (size A) or preferably a P.C.L. suspended-level dilution viscometer*, thermostat bath, electric stirrer.

MATERIALS

Benzoyl peroxide, styrene, A.R. chloroform, methyl alcohol, potassium hydroxide.

DISCUSSION

The polymerisation of styrene (vinyl benzene) is an example of vinyl or addition polymerisation where n molecules of a monomer A unite to produce a polymer $[A]_n$ without the elimination of any molecules from the system. In amide or ester condensation, on the other hand, macromolecular formation occurs only with the elimination of water from the reactants (*see* page 165).

Vinyl polymerisation is generally initiated by a substance which decomposes to a free radical R˙ containing an unshared single electron

$$R^{\bullet} + \overset{H}{\underset{H}{C}} :: \overset{H}{\underset{Ph}{C}} \;\rightarrow\; R : \overset{H}{\underset{H}{C}} : \overset{H}{\underset{Ph}{C}}{}^{\bullet}$$

* Obtainable from A. D. Whitehead (Polymer Consultants Ltd.) 'Ancient House', Ardleigh. nr. Colchester, England.

$$R:\overset{\overset{H}{\cdot\cdot}}{\underset{\underset{H}{\cdot\cdot}}{C}}:\overset{\overset{H}{\cdot\cdot}}{\underset{\underset{Ph}{\cdot\cdot}}{C}}\cdot \; + \; \overset{\overset{H}{\cdot\cdot}}{\underset{\underset{H}{\cdot\cdot}}{C}}::\overset{\overset{H}{\cdot\cdot}}{\underset{\underset{Ph}{\cdot\cdot}}{C}} \;\rightarrow\; R:\overset{\overset{H}{\cdot\cdot}}{\underset{\underset{H}{\cdot\cdot}}{C}}:\overset{\overset{H}{\cdot\cdot}}{\underset{\underset{Ph}{\cdot\cdot}}{C}}:\overset{\overset{H}{\cdot\cdot}}{\underset{\underset{H}{\cdot\cdot}}{C}}:\overset{\overset{H}{\cdot\cdot}}{\underset{\underset{Ph}{\cdot\cdot}}{C}}\cdot \; \text{etc.}$$

or

$$RCH_2CHPhCH_2\overset{\cdot}{C}HPh$$

A commonly used initiator is dibenzoyl peroxide which probably produces free radicals by the following mechanism

$$(RCOO)_2 \rightarrow 2R\cdot + 2CO_2$$

Polystyrene, widely used in the plastic field, finds some application also in textiles in the form of fibres and bristles.

EXPERIMENTAL PROCEDURE

(a) *Polymerisation of styrene*—Purify the benzoyl peroxide (2 g) by dissolving in cold chloroform (5 ml) and· adding methyl alcohol (5 ml). Allow to stand, filter the solid on a small sintered glass funnel, wash with methyl alcohol and air-dry. Extract styrene (100 ml) with two portions (20 ml each) of 10 per cent sodium hydroxide solution in a separating funnel to remove the inhibitor. Extract with two portions of water (20 ml each), dry the styrene overnight over solid potassium hydroxide and filter.

Take four stoppered Quickfit test-tubes (labelled) and add 0·005, 0·010, 0·050 and 0·100 g of purified benzoyl peroxide to individual tubes. Add 10·0 ml of inhibitor-free styrene to each and shake to dissolve the initiator. Allow the tubes to stand in a water bath at 80°C for 2 h when the contents should appear viscous. Cool the tubes and transfer their contents to small beakers by the use of benzene (25 ml) as solvent. Add the benzene solution dropwise from a small separating funnel to methyl alcohol (200 ml) in a 400 ml beaker with vigorous mechanical stirring. Filter the precipitated samples of polystyrene on a Buchner funnel with the aid of a water-pump, wash with two 25 ml portions of methyl alcohol and dry in a vacuum over silica gel. Record the yields.

(b) *Determination of viscosity-average molecular weights*—Dry five portions (0·1, 0·2, 0·3, 0·4 and 0·5 g) of each of the polystyrene samples over phosphorus pentoxide and weigh accurately. Dissolve each in A.R. chloroform in a 100 ml graduated flask.

Filter the solutions through glass wool and determine the relative viscosity (η_{rel}) at 25°C against chloroform alone using a size A Ostwald viscometer. (*See* page 168).

Then,

$$\eta_{rel} = \frac{\text{Flow time of solution}}{\text{Flow time of solvent}} \text{ and } \eta_{sp} = \eta_{rel-1}$$

The viscosity number (reduced viscosity) $= \eta_{sp}/c$

where $c = $ concentration in g/100 ml

Plot viscosity number against c for each of the four polystyrene samples, the intercepts (on the y axis) giving the limiting viscosity number (intrinsic viscosity), $[\eta]$. *See* Figure 24.

Calculate the viscosity-average molecular weight $\overline{M}_v$ of each of the polystyrene samples from its limiting viscosity number using the equation

$$[\eta] = k\overline{M}_v^a$$

where, under the experimental conditions used,

$$k = 1.12 \times 10^{-4} \text{ and } a = 0.73$$

Check the dependence of molecular weight on catalyst concentration by plotting a graph of $\log \overline{M}_v$ against $\log$ benzoyl peroxide concentration in g.moles/l and measure the slope.

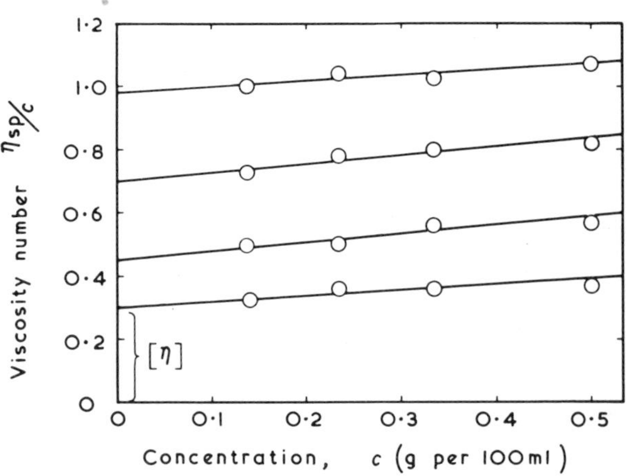

Figure 24. Plots of viscosity number v. concentration of polystyrene for different concentrations of initiator

Since $\bar{P} = k'M/B^{\frac{1}{2}}$, where $\bar{P} =$ average degree of polymerisation, $B =$ benzoyl peroxide concentration (moles/1 of styrene) and $M =$ monomer concentration (moles/1).

Then $\bar{M}_v \propto B^{-\frac{1}{2}}, \bar{M}_v = k''B^{-\frac{1}{2}}$ and $\log \bar{M}_v = -\frac{1}{2}\log B + C$, where C is a constant.

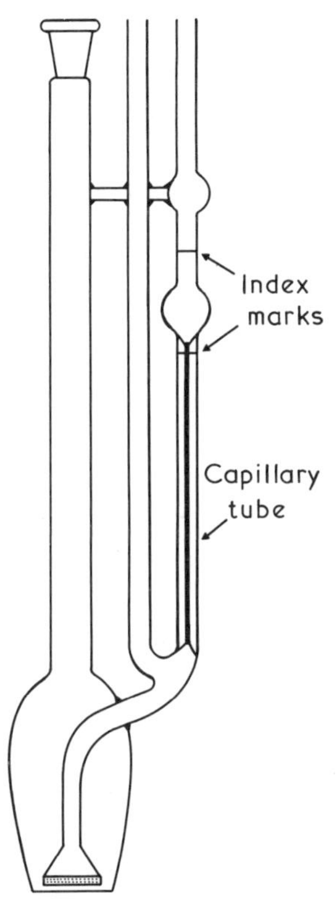

Figure 25. Suspended level dilution viscometer

(*c*) *Use of the suspended level dilution viscometer*—The determination of the viscosities of a number of polymer solutions differing in concentration is tedious when the Ostwald viscometer is used as described under (b) and considerable time may be saved by using a suspended level dilution viscometer. It is satisfactory to use the P.C.L. standard model (Fig. 25) with a capillary diameter of 0·4 mm which requires a minimum of 4·5 ml and a maximum of 40·0 ml of liquid for testing. It is adapted from the Ubbelohde viscometer and is operated as follows (*see* P.C.T. Bulletin [1]):

The viscometer is clamped vertically in a thermostatic bath. The solvent is introduced into the conical bulb of the viscometer by pipette, the volume used being unimportant provided it lies within the limits 4–40 ml. The air tube in the centre is closed with a rubber policeman or with the finger and, by means of a rubber bulb fixed to the filling tube, the liquid is forced up the capillary until the uppermost bulb is half-filled. Caution must be exercised at this stage to avoid blowing the liquid out of the air tube. This tube must be kept firmly closed and only moderate pressure should be used to force the liquid through the capillary. On releasing the pressure and venting the air tube, the liquid level immediately breaks in the lower bulb, thus forming the suspended level. The flow of the solvent between the upper and lower index marks may now be timed. After repeating this process two or three times, the viscosity/ density ratio may be calculated from the mean flow time using the calibration constants.

For determining limiting viscosity numbers, the calibration constants need not be known and only one solution of the polymer in a suitable solvent (0·5–1·0 per cent) need be prepared. The flow time for the solvent is determined first as described above, after which the viscometer is thoroughly drained and dried. With a pipette, 5·0 ml of the polymer solution is delivered into the conical bulb, taking care not to wet the walls of the filling tube, and the flow time of the solution is determined as described. By means of a pipette, 5·0 ml of solvent is now added to the conical bulb. The rubber bulb is attached to the capillary tube and, with the air tube stoppered, a gentle stream of air is blown through the solution to ensure thorough mixing. The flow time is again determined and the mixing is repeated until constant flow times are obtained. The mean flow time is recorded for this dilution and additional portions of 5·0 ml of solvent may be added, mixed in the same way, and the mean flow time determined for each dilution. Five to seven values may be obtained, allowing extrapolation to infinite dilution and evaluation of the limiting viscosity number and the viscosity/concentration relation. If the minimum value of 4·0 ml·of solution is added initially, tenfold dilution may be achieved in the viscometer. For reproducible results, it is essential to keep solution and solvent free from dust.

REFERENCE
1 Polymer Consultants Technical Bulletin, Colchester, England. No. 2V (1962)

46 Polymerisation of acrylonitrile

APPARATUS

A 100 ml three-necked round-bottom flask, electric stirrer, Size A Ostwald viscometer and small hypodermic syringe.

MATERIALS

Acrylonitrile, α,α'-azoisobutyronitrile (azo-bis-isobutyronitrile, dimethylformamide, sodium thiocyanate, silica gel (chromatography grade, approximately 60–120 mesh), nitrogen-cylinder (*see* Note).

DISCUSSION

In the experimental procedure described, acrylonitrile is polymerised in homogeneous solution

$$
n \begin{array}{c} H \\ | \\ C \\ | \\ H \end{array}\!\!=\!\!\begin{array}{c} CN \\ | \\ C \\ | \\ H \end{array} \rightarrow -\!\!\begin{array}{c} H \\ | \\ C \\ | \\ H \end{array}\!\!-\!\!\begin{array}{c} CN \\ | \\ C \\ | \\ H \end{array}\!\!-\!\!\begin{array}{c} H \\ | \\ C \\ | \\ H \end{array}\!\!-\!\!\begin{array}{c} CN \\ | \\ C \\ | \\ H \end{array}\!\!-\!\!\begin{array}{c} H \\ | \\ C \\ | \\ H \end{array}\!\!-\!\!\begin{array}{c} CN \\ | \\ C \\ | \\ H \end{array}\!\!-
$$

The polymerisation takes place via a free-radical mechanism (*see* page 156). Sodium thiocyanate solution is used as solvent because its low transfer constant produces a polymer of high molecular weight (*see* Billmeyer [1]).

EXPERIMENTAL PROCEDURE

WARNING Acrylonitrile vapour is highly toxic and the liquid (BP 76°C) must be handled in a fume cupboard. Dimethylformamide must not be allowed to come into contact with the skin and breathing the vapour must be avoided.

(*a*) *Polymerisation of acrylonitrile*—Commercial acrylonitrile contains an inhibitor (usually a polyhydric phenol) which is removed by passing the acrylonitrile through a column of dry silica gel immediately before use. A 30 cm × 2 cm column will purify 100 g of acrylonitrile. Once the inhibitor has been removed, the acrylonitrile may polymerise spontaneously, particularly in sunlight. The silica gel may be regenerated by passing 2N-sodium hydroxide (500 ml) through the column followed by water (1 litre) drying and repacking.

Dissolve sodium thiocyanate (20 g) in freshly boiled distilled water (20 ml). To this solution in a 100 ml three-necked round-bottom flask add inhibitor-free acrylonitrile (6·0 g or 7·5 ml). Fit the flask with a single-surface water condenser, stirrer and thermometer. The bulb of the latter is immersed in the solution and must not be fouled by the stirrer. A stream of nitrogen is introduced into the flask by means of a glass tube inserted through the condenser and terminating above the surface of the liquid (*see* Note). The flask is immersed in a water bath at 70°C. When the contents have attained bath temperature, add α,α′-azoisobutyronitrile (0·010 g) previously dissolved by warming in a few drops of the thiocyanate solution. Within a few minutes of adding the α,α′-azoisobutyronitrile the temperature inside the flask rises to 73–74°C. Maintain the bath temperature at 70°C for 30 min after the addition of the initiator. During the initial stages of the preparation, stirring should be fairly rapid to effect homogeneous solution and thermal equilibration of the contents. When this has been achieved, stirring need not be as rapid.

(*b*) *Conversion of polymer to film*—The polymer solution prepared as described under (a) must be converted to films immediately since it sets to a solid gel on cooling. The concentration of the dope is determined by casting a known weight into films. The dope is poured on to a glass plate (a 10 cm × 10 cm T.L.C. plate is very suitable) until an accurately known weight increase of approximately 3 g is obtained. A similar glass plate is pressed on top and when the dope has spread between the plates, the two are slid apart to leave a thin even film on each. Immerse the plates in a beaker of water at 50–60°C for about 5 min. Peel off the films and wash them successively in boiling 0·1N-hydrochloric acid (250 ml) for 10 min and boiling distilled water for 15 min. Dry at 60°C or in a vacuum desiccator over silica gel and weigh. Determine the dope concentration and hence the percentage conversion. (If x g of dope give y g of film, the percentage conversion is $4600y/6x$. Retain a few millilitres of dope for extrusion (*see* d(iii)) and convert the remainder to films which are required for the following experiments.

(*c*) *The viscosity of solutions of polyacrylonitrile*—Keeping all other variables constant, including the time, repeat the polymerisation procedure described using 0·020 g, 0·050 g and 0·100 g of initiator. Convert each solution to film and determine the percentage yield of polymer in each case.

Measure the relative viscosity (*see* page 157) of a 0·50 per cent solution (0·250 g per 50 ml) of each of the polyacrylonitrile samples in dimethylformamide (DMF) at 25°C using a size A Ostwald viscometer. Express the results as inherent viscosity, η_{inh} (*see* (Sorenson and Campbell [2])

$$\eta_{inh} = \ln \frac{\eta_{rel}}{c}$$

where c = concentration of solution in g per 100 ml.

Comment on the effect of initiator concentration on the percentage conversion and the molecular weight of the polymer produced. The inherent viscosity is a measure of the latter.

(*d*) *Properties of polyacrylonitrile*
   (i) Examine the effects of common solvents for polymers, for example acetone, chloroform, benzene, concentrated formic acid, etc., on small portions of film and comment on the results.
   (ii) Heat some film on a watch glass in an oven or furnace at 250–275°C for 4–6 h. Compare the properties of the resulting 'carbon' film with those of the original, particularly with regard to flammability.
   (iii) Extrude a small quantity of the thiocyanate dope through a hypodermic syringe using a No. 14 needle into water at 50–60°C. Attempt to extend the filaments as they are extruded. Wash and dry the fibres and comment on them.

163

M

NOTE

Although it is possible to prepare a polymer without the use of nitrogen, it is preferable that the polymerisation be performed in an atmosphere of nitrogen if it is intended to correlate viscosity with the concentration of initiator. It is convenient to pass the gas from the cylinder through a water-bubbler to obtain an indication of the flow-rate. The nitrogen should be passed rapidly for a few minutes to flush the oxygen from the flask and then maintained throughout the polymerisation at a rate of a few bubbles per minute.

REFERENCES
1 BILLMEYER JR., F. W., *Textbook of Polymer Science*. Interscience Publishers Inc., New York, 266 (1962)
2 SORENSON, W. R. and CAMPBELL, T. W., *Preparative Methods of Polymer Chemistry*. Interscience Publishers Inc., New York, 35 (1962)

47 Preparation of nylon 66 and nylon 610

APPARATUS

Hard glass reaction tube, No. 2 sintered filter funnel (small), melting point apparatus, B.S.S. Ostwald viscometer (size A) and a thermostat bath.

MATERIALS

Adipic and/or sebacic acid, hexamethylenediamine, ethyl alcohol (industrial methylated spirit), 90 per cent w/w formic acid (A.R.), Wood's metal, glass wool and, optionally, nitrogen cylinder.

THEORY

Under suitable conditions dicarboxylic acids and diamines condense to form long-chain polyamides.

$$H_2N[CH_2]_xNH_2 + HOOC[CH_2]_yCOOH + H_2N[CH_2]_xNH_2$$
$$+ HOOC[CH_2]_yCOOH$$
$$\downarrow$$
$$-HN[CH_2]_xNHCO[CH_2]_yCONH[CH_2]_xNHCO[CH_2]_yCO-$$
$$+ nH_2O$$

When the chains are of suitable length, these polyamides may be melted and extruded into filaments, the nylon 66 of commerce being made from hexamethylenediamine and adipic acid ($x = 6$ and

$y = 4$). The different types available and the nomenclature used to describe them are given by Gordon Cook [1] and Peters [2].

EXPERIMENTAL PROCEDURE

(*a*) *Preparation of nylon 66—polyhexamethyleneadipamide*—Dissolve adipic acid (3·6 g or 0·024 mole) in ethyl alcohol (30 ml) and dissolve hexamethylenediamine (2·9 g or 0·025 mole) in a mixture of ethyl alcohol (8 ml) and distilled water (3 ml). Add the diamine solution to the adipic acid solution with stirring. The neutralisation is exothermic and the nylon salt precipitates on cooling. The solution should be allowed to stand at room temperature, or preferably lower, for at least 30 min to ensure that precipitation is as complete as possible. Filter the white crystalline nylon salt on a Buchner funnel, wash twice with alcohol and dry at 110°C or in an evacuated desiccator over silica gel. Determine its melting point (196°C).

Place an intimate mixture of nylon salt (2·5 g) and adipic acid (0·028 g) in a hard glass reaction tube (Figure 26). Partly draw out

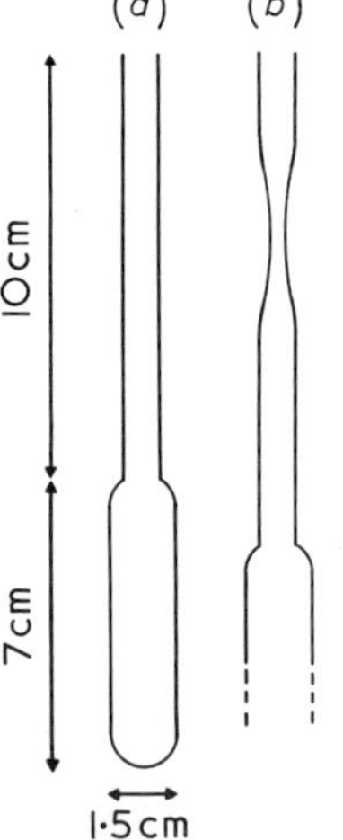

Figure 26. Reaction tube (a) and partly drawn out (b)

the tube and pass nitrogen in through a capillary tube for 5–10 min. Remove the capillary and seal off the tube. (Although flushing with nitrogen is desirable, if nitrogen is not available this step may be omitted.) Place in an oven at 220°C for 2 h enclosed in a tin. Allow the tube to cool, cut off the top neatly and connect by pressure tubing to a water pump (*see* page   ). Heat the tube under vacuum in a bath of Wood's metal at 270°C for 2 h to complete the condensation.

Remove the tube from the bath and allow to cool whilst still under vacuum. (The tube may splinter on cooling.) Place the tube in a cloth and break it open to remove the translucent plug of nylon 66. By means of a hot wire, pluck about 20 filaments from the polymer. Extend half of these by hand and note that a point is reached where the fibre does not yield readily but has the elasticity of a true textile fibre (*see* Note 1).

The nylon 66 is purified by dissolving with shaking in 90 per cent formic acid (25 ml) and filtering through a small No. 2 sintered filter funnel into distilled water (200 ml). The precipitated nylon is collected in a similar filter funnel, washed free of acid with distilled water and dried in air.

(*b*) *Preparation of nylon 610—polyhexamethylenesebacamide* ($x = 6$, $y = 8$)—Nylon 610 salt is prepared as described under (a) from sebacic acid (5·0 g) and hexamethylenediamine (2·9 g). Record the melting point (170°C).

A mixture of nylon 610 salt (3·0 g) and sebacic acid (0·039 g) is converted to the condensation polymer as described under (a) but the heating at 270°C is carried out for 3 h. It is purified by precipitation from formic acid as before.

(*c*) *Determination of molecular weights* (Taylor [3])—Dry the polymer (approximately 0·5 g) in a vacuum desiccator over phosphorus pentoxide, weigh accurately, dissolve in 90 per cent w/w formic acid and make up to 100 ml in a graduated flask. Filter the solution through glass wool and determine the viscosity ratio (η_{rel}) at 25°C against 90 per cent w/w formic acid alone, using a size A B.S.S. Ostwald viscometer (*see* Note 2).

$$\eta_{rel} = \frac{\text{Time of flow of solution}}{\text{Time of flow of formic acid solvent}}$$

and

$$\eta_{sp.} = \eta_{rel} - 1 \text{ where } \eta_{sp} = \text{specific viscosity}$$

The intrinsic viscosity (limiting viscosity number) is calculated from the specific viscosity by means of the Huggins–Kraemer equation.

$$\frac{\eta_{sp}}{c} = [\eta] + k^1 [\eta]^2 c$$

where c = concentration in g/100 ml, $[\eta]$ = the intrinsic viscosity and $k^1 = 0.11$.

The number-average molecular weight $\overline{M}_n$ is finally calculated from the equation

$$\overline{M}_n = k[\eta]^\alpha$$

where $k = 13000$ and $\alpha = 1.39$.

The values for k^1, k and α have been found for nylon 66 in 90 per cent formic acid at 25°C and with this polymer an accurate value for $\overline{M}_n$ may be obtained using the constants given. The same constants, however, will give an approximate value for $\overline{M}_n$ if applied to polymer 610.

NOTES

1. Information regarding the orientation of the molecular chains in these filaments in the drawn and undrawn states may be obtained from their birefringence, x-ray diffraction patterns or polarised infra-red absorption spectra (*see* Peters [4]).

2. To obtain accurate results, the following must be observed.

The viscometer is cleaned with a mixture of concentrated sulphuric acid and sodium or potassium dichromate (CAUTION), washed well with tap water and finally distilled water. It is dried in an oven at 60–80°C. Other glassware and solutions must be completely free from foreign particles.

The viscometer is filled exactly to the mark just above the lower reservoir bulb (these instructions apply to Ostwald viscometers). Since viscosity is very sensitive to temperature, all measurements should be made with the viscometer in a thermostat bath held constant within ± 0.05°C. It is good practice to keep the flasks of solvent and solution in the thermostat bath and to fill the viscometer whilst immersed in the bath. The viscometer must be supported rigidly in a vertical position and time allowed for the whole to reach thermal equilibrium.

By means of a rubber bulb attached to the filling tube, force liquid up the capillary tube, filling the small bulb to above the upper mark. Remove the rubber bulb and allow the liquid to fall freely. Using a stop-watch, determine the time for the meniscus to pass from the upper to the lower mark. Repeat flow-times should be reproducible to within 0.2 s if the solutions are free from dust and the temperature is constant. Under the experimental conditions given, flow-times will be considerably greater than 100 s and no kinetic energy correction need be applied. The same viscometer must be used for solvent and solution.

Although for accurate work it is essential to use a thermostat bath, approximate values for these molecular weights may be obtained by determining the times of flow with the viscometer immersed in a large beaker of water with the temperature kept constant $\pm 0.5°C$ by means of a small bunsen burner flame. By this means a molecular weight of 14 000 should have an error of less than ± 1000.

REFERENCES
1 GORDON COOK, J., *Handbook of Textile Fibres*. vol. II. Merrow, Watford, England, 212 (1968)
2 PETERS, R. H., *Textile Chemistry*. vol. I. Elsevier, Amsterdam, 27 (1963)
3 TAYLOR, G. B., *J. Am. chem. Soc.,* **69**, 635 (1947)
4 PETERS, R. H., *Textile Chemistry*, vol. I. Elsevier, Amsterdam, 393 (1963)

48 Identification of the hydrolysis products of nylon 66

APPARATUS

Small-scale organic preparative apparatus, melting-point apparatus.

MATERIALS

Industrial methylated spirit (Ethyl alcohol), benzoyl chloride, nylon 66 (preferably as bright yarn), 0·10N-sodium hydroxide.

THEORY

Nylons, in general, are made by condensing dicarboxylic acids with diamines (*see* Gordon Cook [1] and Moncrieff [2]). Under severe hydrolytic conditions, the reaction is reversed and nylons yield the corresponding acids and diamines:

$$-NH[CH_2]_xNHCO[CH_2]_yCO-$$
$$\downarrow H_2O$$
$$H_2N[CH_2]_xNH_2 + HOOC[CH_2]_yCOOH$$

In the case of nylon 66 or polyhexamethyleneadipamide, the products formed are adipic acid, $HOOC[CH_2]_4COOH$, and hexamethylenediamine, $H_2N[CH_2]_6NH_2$. The characterisation of these products may be used to determine the structure of a particular nylon.

Adipic acid may be identified by its melting point and equivalent weight, but hexamethylenediamine is not readily purified and is best characterised by the melting point of a suitable derivative, such as the N, N^1 dibenzoyl, prepared using benzoyl chloride:

$$H_2N[CH_2]_6NH_2 + 2C_6H_5COCl \rightarrow$$
$$\rightarrow C_6H_5CONH[CH_2]_6NHCOC_6H_5 + 2HCl$$

EXPERIMENTAL PROCEDURE

Although nylon yarn may sometimes be used without purification, to ensure that it does not contain an excessive amount of oil it should be washed in a beaker with cold carbon tetrachloride and air dried.

Dilute 10·0 ml of conc. hydrochloric acid with 10·0 ml of distilled water to produce approximately 6N-acid. In a drawn out soft soda glass test-tube, place 0·52 g of nylon 66 yarn, weighed after standing in the atmosphere. Pour in 10·0 ml of 6N-hydrochloric acid, seal off and shake until the nylon is dispersed. Prepare a second tube in an identical manner. The material to be hydrolysed corresponds to 1·00 g dry weight, and the hydrolysis is performed in two tubes as a safety measure. Place in an oven at 120°C for 5 h and leave to cool overnight in the oven (*see* page ix).

Holding the tube in a cloth, break off the constricted part with pliers and transfer the whole of the contents of the two tubes into a small conical flask. The hydrolysate should contain crystalline adipic acid. Stand the flask in the refrigerator for 2 h to complete the crystallisation and filter through a small Hirsch funnel.
Crystalline precipitate—Transfer to a small glass dish and dry in a vacuum desiccator over solid caustic soda. Record (a) the total weight (b) the calculated yield and (c) the yield as a percentage of the theoretical. Retain a few milligrammes of the product for identification by paper chromatography (*see* page 176).

Make up aq. ethyl alcohol by mixing water (15 ml) and industrial methylated spirit (5 ml). Dissolve the crude adipic acid in 5 ml of this solvent using gentle heat. Filter under suction and stand the filtrate in the refrigerator for 15 min. Filter and wash the product with a few millilitres of cold solvent. A further crop may be obtained by allowing the filtrate to stand in the refrigerator overnight. Record (d) the melting point of the purified adipic acid (150–152°C) and (e) its equivalent weight. Calculate (f) the theoretical equivalent weight.

To determine the equivalent weight, weigh out accurately approximately 0·1 g of the product, dissolve in approximately 20 ml of cold water, add phenolphthalein indicator and titrate to a pink colour with 0·10N-NaOH. If the volume of the latter is V ml,

$$\text{Equivalent weight} = \frac{10000w}{V}$$

where w = wt. of acid (g).

Filtrate—Place in a glass dish and take down to dryness in a vacuum desiccator over solid caustic soda. Take approximately half the product, retaining the remainder for examination by paper chromatography, and dissolve in 10 per cent aq. sodium hydroxide solution (20 ml). Add benzoyl chloride (1·5 ml) with shaking over a period of 10 min. Filter off the crude NN^1-dibenzoyl derivative under vacuum and wash with cold water until the washings are free from alkali. Mix industrial methylated spirit (60 ml) and water (40 ml) and dissolve the product in 50 ml of this with gentle heat. Filter under vacuum and allow the filtrate to cool. Filter off the crystalline precipitate and wash with a small quantity of the same cold aq. alcohol. Dry on filter paper and finally in an oven for 10 min at 110°C. Record (g) the melting point (159–161°C).

REFERENCES
1 GORDON COOK, J., *Handbook of Textile Fibres*. vol. II. Merrow, Watford, England, 212 (1968)
2 MONCRIEFF, R. W., *Man-made Fibres*. Heywood, London, 299 (1963)

49 Identification of the hydrolysis product of nylon 6

APPARATUS

Apparatus required to undertake small-scale organic preparative work, melting points and Kjeldahl nitrogen determinations (*see* page 23).

MATERIALS

Nylon 6 (preferably as bright yarn), benzyl chloroformate (handle with care), ether, acetone, decolorising charcoal, 0.10N-sodium hydroxide, material for Kjeldahl nitrogen determination (*see* page 23) sodium selenate.

THEORY

Nylon 6 (often known as Perlon L) is made by the polymerisation of caprolactam.

$$n \underset{NH}{\overset{CH_2[CH_2]_4CO}{\diagdown \diagup}} \rightarrow -NH[CH_2]_5CONH[CH_2]_5CO-$$

and on hydrolysis this yields 6-aminocaproic acid

$$-NH[CH_2]_5CONH[CH_2]_5CO \xrightarrow{H_2O} n\, H_2N[CH_2]_5COOH$$

This may be identified by paper chromatography or more positively by formation of the carbobenzoxy derivative using benzyl chloroformate.

$$C_6H_5CH_2OCOCl + H_2N[CH_2]_5COOH$$

$$\xrightarrow{NaOH} C_6H_5CH_2OCONH[CH_2]_5COOH + HCl$$

The latter may be characterised by means of its melting point, equivalent weight and nitrogen content. Carbobenzoxy derivatives of amino acids are used widely in the first stage (N-protection) of many peptide syntheses (*see* Bergman and Zervas [2] and Bailey [3]) since it is necessary to protect the N-terminal amino group before coupling one amino acid with another.

EXPERIMENTAL PROCEDURE

The nylon 6 is washed in cold carbon tetrachloride to remove oil and then air-dried.

Air-dried nylon 6 (1·04 g) is heated under gentle reflux, using a water condenser, for 8 h with a mixture of conc. hydrochloric acid (10·0 ml) and water (10·0 ml).

The hydrolysate is taken down to dryness in a vacuum desiccator over solid sodium hydroxide and a few milligrammes of the resulting aminocaproic acid hydrochloride retained for investigation by paper chromatography (*see* page 176).

Dissolve the remainder of the residue in 10 per cent sodium hydroxide solution (10 ml) and transfer to a 100 ml conical flask with a glass stopper. Cool to 0°C in a mixture of ice and water, and over a period of 15 min, add 20 per cent sodium hydroxide solution (3·0 ml) and benzyl chloroformate (3·0 ml) with alternate vigorous shaking and cooling. Operations involving benzyl chloroformate should be performed in a fume cupboard. In each case, add the sodium hydroxide before the benzyl chloroformate to ensure that the reaction mixture remains alkaline. Allow the mixture to stand at 0°C for 15 min. Extract with ether (20 ml) in a separating funnel, reject the ethereal layer and acidify the clear aq. solution to pH 1 with conc. hydrochloric acid, as shown by a wide-range test paper. Allow the flask to stand in the refrigerator until the separated oil solidifies. Filter under vacuum from a water pump and wash the derivative very thoroughly with ice-cold water. Dry under vacuum over solid sodium hydroxide. Calculate (a) the theoretical yield and record (b) the actual yield in grammes and (c) as a percentage of the theoretical.

To purify the crude derivative, it is dissolved in a mixture of acetone (15 ml) and water (30 ml) with gentle heat. About a gramme of decolorising charcoal is added, followed by stirring for a few minutes. The warm solution is filtered through a Buchner or Hirsch funnel under vacuum and allowed to stand in the refrigerator. The white solid is filtered under vacuum, washed with a small volume of cold solvent (2:1 water–acetone) and dried under vacuum in a desiccator containing sodium hydroxide pellets. Record (d) the melting point of the derivative (52°C), (e) its nitrogen content and (f), its equivalent weight. Calculate the theoretical values for (e) and (f).

The nitrogen content may be determined by the Kjeldahl method but the carbobenzoxy derivative does not digest too readily and it is preferable to replace the copper sulphate catalyst by sodium selenate (0·2 g).

To determine the equivalent weight, weigh out accurately approximately 0·2 g of the product, dissolve in a mixture of acetone (10 ml) and water (10 ml), add phenolphthalein indicator and titrate to a pink colour with 0·10N-sodium hydroxide (for calculation *see* page 172).

REFERENCES
1 MONCRIEFF, R. W., *Man-made Fibres.* Heywood, London, 335 (1963)
2 BERGMANN, M. and ZERVAS, L., *Ber.,* **65**, 1192 (1932)
3 BAILEY, J. L., *Techniques in Protein Chemistry.* Elsevier, Amsterdam, 353 (1967)

50 Chromatographic identification of nylons

APPARATUS

Shandon 'Unikits' Nos. 1 and 2 including 10 cm × 10 cm plates.

MATERIALS

Nylon 6 (Perlon L or 'Celon'), nylon 66 ('Bri-Nylon'), nylon 610 if available. Adipic acid, $HOOC(CH_2)_4COOH$, suberic acid, $HOOC(CH_2)_6COOH$, sebacic acid, $HOOC(CH_2)_8COOH$, 6-amino-n-hexoic acid (ε-amino-n-caproic acid), $H_2N(CH_2)_5COOH$ and hexamethylenediamine, $H_2N(CH_2)_6NH_2$. Bromocresol green, ethyl alcohol, n-propanol, Whatman No. 1 paper (25 cm × 25 cm), Kieselgel G, Whatman CC41 cellulose powder.

DISCUSSION

On hydrolysis, nylons break down into dibasic acids and diamines or they give rise to ω-aminocarboxylic acids (*see* page 170 and 174). Thus, nylon 66 produces hexamethylenediamine and adipic acid, 610 produces hexamethylenediamine and sebacic acid, whilst nylon 6 gives one substance only, 6-amino-n-hexoic acid. These products may be readily identified by chromatography using paper or thin-layer plates.

EXPERIMENTAL PROCEDURE

Hydrolyse the nylon with 5N-hydrochloric acid (*see* page 183 or 186) and take down to dryness over solid sodium hydroxide (*see* page 174. Dissolve the hydrolysate (approximately 0·05 g) in water (10 ml). Dissolve also the reference substances in water to give solutions of the same concentration.

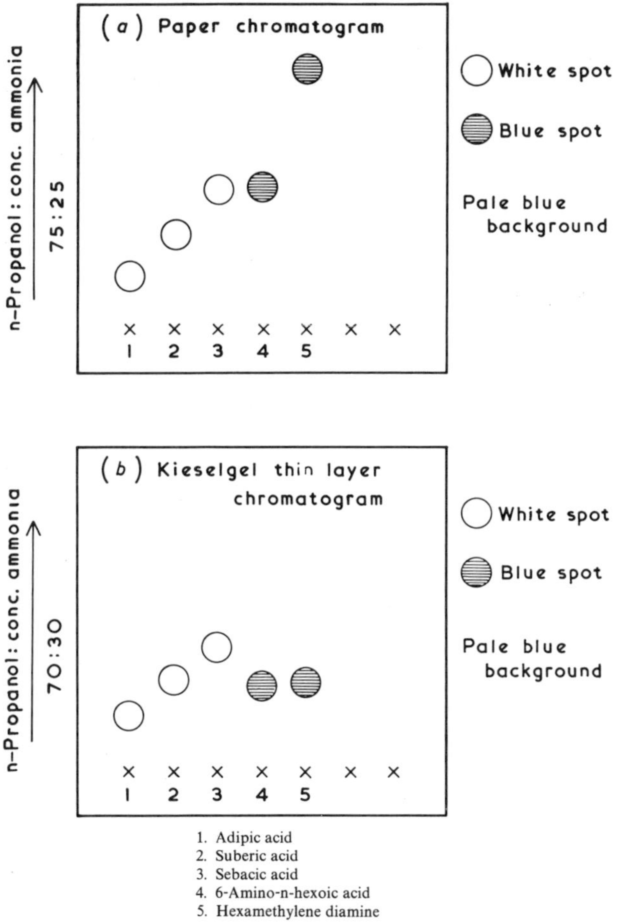

1. Adipic acid
2. Suberic acid
3. Sebacic acid
4. 6-Amino-n-hexoic acid
5. Hexamethylene diamine

Figure 27. Separation of nylon hydrolysis products by one-way chromatography

177

The locating spray is prepared as follows: 0·05N-sodium hydroxide is prepared by dissolving sodium hydroxide (2 g) in water (1 litre). Bromocresol green (0·1 g) is dissolved in a mixture of 0·05N-sodium hydroxide (2·9 ml) and 90 per cent alcohol (5 ml) with warming. The solution is made up to 250 ml with 20 per cent alcohol to give a final concentration of 0·04 per cent.

Make up two solvents for running the chromatograms consisting of (a) n-propanol (75 ml) and 0·880 ammonia solution (25 ml) and (b) n-propanol (70 ml) and 0·880 ammonia solution (30 ml). Run the hydrolysates on a chromatogram together with the reference substances bearing in mind that the locating spray gives white spots with carboxylic acids, but blue spots with diamines and aminocarboxylic acids. The background is pale blue. Good separations may be achieved by running the chromatograms in one direction (ascending) only, using the following support media and solvents (*see* Figure 27).

(i)   Whatman No. 1 paper and solvent (a)
(ii) Kieselgel thin layer plates and solvent (b)
(iii) Whatman CC41 cellulose thin layer plates and solvent (a)

Comment on the position of the dicarboxylic acid spots and the number of carbon atoms in the molecule.

REFERENCE
1 RAVEN, D. J. and EARLAND, C., *J. Soc. Dyers. Cols.*, **86**, 313 (1970)

51 Properties of polyester fibres

APPARATUS

Small sintered glass funnels (Nos. 1 and 2), B.S. 188 Ostwald viscometer (size B), thermostat bath electric stirrer.

MATERIALS

Terylene or Dacron yarn or staple (bright or delustred), 0·10N-
sodium hydroxide, thymol blue indicator (B.P.), ethyl alcohol,
methyl cellosolve (ethylene glycol monomethyl ether), *o*-chloro-
phenol.

Warning—*o*-Chlorophenol is highly corrosive and toxic.

DISCUSSION

Polyesters suitable for extrusion into fibres may be prepared by the
condensation of suitable dicarboxylic acids with diols. At the
present time the market is dominated by Terylene (I.C.I. Ltd., U.K.)
and Dacron (du Pont, U.S.A.) which are both polyethylene terephthalate fibres derived from terephthalic acid and ethylene glycol.

The industrial method of manufacture of Terylene involves an
ester-interchange reaction (*see* Gordon Cook [1]).

The only other commercial polyester fibre is Kodel (Eastman
Chemical Products, Inc, U.S.A.) which is a polyester with the same
acid constituent. Since under severe hydrolytic conditions polyesters
break down into their component acids and alcohols, this class of

$$-OCH_2CH_2OOC\langle\ \rangle CO- + nH_2O$$

Polyethylene terephthalate

$$HOOC\langle\ \rangle COOH + HOCH_2CH_2OH$$

Terephthalic acid

fibre may be identified by the isolation of terephthalic acid from a hydrolysate.

As with polyamide fibres, chemical degradation of polyesters, resulting in a shortening of their molecular chains, may be conveniently measured by the intrinsic viscosity of the material in a suitable solvent.

EXPERIMENTAL PROCEDURE

(a) *Isolation and characterisation of terephthalic acid* (*see* Note 1)— In a 250 ml round-bottomed flask fitted with a water reflux condenser, gently boil the polyester yarn or staple (2·00 g) with water (50 ml) and sodium hydroxide (10·0 g) for 4 h. The source of heat may be a small bunsen burner flame or an electric mantle. The addition of a small piece of porous pot reduces bumping. Add water (100 ml) down the condenser, heat for a further period of a few minutes and filter either through a fluted paper or a No. 2 sintered glass funnel under vacuum from a water pump. This will remove any titanium dioxide which may have been present in the original material as a delustrant.

Cool the clear filtrate and, with stirring, acidify with concentrated hydrochloric acid until the solution is below pH 1 (test paper). Cool and filter the terephthalic acid onto a paper in a small Buchner funnel, using a vacuum if necessary. Wash the product with hot water (1 litre), transfer to a small weighed dish, dry at 110°C, re-weigh and record the yield. Express this also as a percentage of the theoretical.

Since terephthalic acid does not possess a well-defined melting point, it is best characterised by determining its equivalent weight. Weigh accurately approx. 0·1 g of the ground product, shake with methyl cellosolve (10 ml), which will not effect complete solution, add 2 spots of thymol blue indicator (see Note 2) and titrate with

0·10N-sodium hydroxide with shaking until the yellow solution changes to blue. Compare the equivalent weight obtained (for calculation see page 172) with the theoretical value.

(b) *Determination of the intrinsic viscosity of Terylene* (*see* Ref. 2)—If necessary, the sample should be freed from spinning oil, etc. by washing in carbon tetrachloride.

The dry Terylene (approx. 0·25 g) is weighed accurately and transferred to a 50 ml round bottomed flask with a ground glass neck. The *o*-chlorophenol (b.p. 175–176°C) is stored in an amber bottle kept in the thermostat and 25·0 ml (at 25°C) is transferred to the flask by means of a bulb operated pipette. The flask which is fitted with an electric stirrer inserted through a gland, is immersed in a boiling water bath. It is essential that no moisture is introduced into the flask. Complete solution is attained within about 30 min. The flask is removed, allowed to cool to 25°C, and the solution is introduced into the viscometer through a No. 1 porosity sintered glass funnel until the liquid is at the correct level in the viscometer. Allow the viscometer and contents to stand in the thermostat bath at 25°C $\pm$ 0·05°C for at least 20 min to attain the correct temperature, check that the solution is at the correct level and determine the time of flow. The timing should be repeated until four consecutive readings agree to within 0·2 sec. In a similar manner determine the time of flow of the *o*-chlorophenol solvent after filtration through the sintered glass funnel.

When determinations on several samples of Terylene are being carried out at one time, it is unnecessary to clean and dry the viscometer between the introduction of successive solutions. After pouring away the old solution, the viscometer should be rinsed twice with portions of the new filtered solution before introducing the main sample; similarly, rinse twice with pure solvent before determining the solvent flow time. The cleaning of the viscometer and other precautions which should be observed in its use are given on page 168.

Results are calculated as follows:

$$\text{Viscosity ratio } (\eta_{rel}) = \frac{\text{Time of flow of solution}}{\text{Time of flow of } o\text{-chlorophenol}}$$

From the table, ascertain the value of the function–'intrinsic viscosity × concentration'–at 25°C corresponding to this viscosity ratio. Let this be x

Table for converting viscosity ratio of a solution containing 1·00 g of Terylene in 100 ml of *o*-chlorophenol to 'intrinsic viscosity × concentration' (*x*)

Viscosity ratio	0·00	0·01	0·02	0·03	0·04	0·05	0·06	0·07	0·08	0·09
1·2	0·189	0·198	0·207	0·216	0·225	0·233	0·242	0·251	0·260	0·268
1·3	0·277	0·285	0·294	0·302	0·311	0·319	0·327	0·336	0·344	0·352
1·4	0·360	0·369	0·377	0·385	0·393	0·401	0·409	0·417	0·425	0·433
1·5	0·441	0·449	0·457	0·464	0·472	0·480	0·488	0·495	0·503	0·511
1·6	0·518	0·526	0·533	0·541	0·548	0·556	0·563	0·571	0·578	0·586
1·7	0·593	0·600	0·608	0·615	0·622	0·629	0·636	0·644	0·651	0·658
1·8	0·665	0·672	0·679	0·686	0·693	0·700	0·707	0·714	0·721	0·728
1·9	0·735	0·742	0·749	0·756	0·763	0·769	0·776	0·783	0·790	0·797
2·0	0·803	0·810	0·817	0·823	0·830	0·837	0·843	0·850	0·856	0·863

Then, Intrinsic viscosity $[\eta] = \dfrac{x}{4y}$

where y = weight of Terylene used.

Determine the intrinsic viscosity of untreated Terylene yarn and the same material which has been heated under reflux on a boiling water bath with equal volumes of concentrated hydrochloric acid and water for 10, 20, 30, 40 and 50 h followed by washing and drying. Represent the results graphically and correlate with the tensile properties of the yarn.

NOTES

1. Before an unknown material is subjected to this procedure, a preliminary identification should be made. It should be established that the fibre is devoid of nitrogen (no ammonia is evolved when it is heated with soda lime) and it is insoluble in acetone, chloroform and concentrated formic acid and sinks in water.

2. Thymol blue indicator solution is made by dissolving the solid (0·10 g) in a mixture of 0·10N-sodium hydroxide (2·2 ml), water (2·2) ml) and 90 per cent (v/v) ethyl alcohol (5·0 ml). After solution is complete, make the volume up to 250 ml with 20 per cent (v/v) ethyl alcohol.

REFERENCES
1 GORDON COOK, J., *Handbook of Textile Fibres*. Vol. II. Merrow, Watford, England, 356 (1968)
2 *I.C.I. Fibres Manual, Section C2, formerly* 'Terylene Technical Information Manual', I.C.I. Fibres Division, Harrogate, Yorks.

Appendix

TABLE OF ATOMIC WEIGHTS FOR USE IN CALCULATIONS

Element	Symbol	Atomic weight	Element	Symbol	Atomic weight
Aluminium	Al	27·0	Mercury	Hg	200·6
Antimony	Sb	121·8	Molybdenum	Mo	96·0
Arsenic	As	74·9	Nickel	Ni	58·7
Barium	Ba	137·4	Nitrogen	N	14·0
Beryllium	Be	9·0	Osmium	Os	190·2
Bismuth	Bi	209·0	Oxygen	O	16·0
Boron	B	10·8	Phosphorus	P	31·0
Bromine	Br	79·9	Platinum	Pt	195·2
Cadmium	Cd	112·4	Potassium	K	39·1
Calcium	Ca	40·1	Selenium	Se	79·0
Carbon	C	12·0	Silicon	Si	28·1
Cerium	Ce	140·1	Silver	Ag	107·9
Chlorine	Cl	35·5	Sodium	Na	23·0
Chromium	Cr	52·0	Strontium	Sr	87·6
Cobalt	Co	58·9	Sulphur	S	32·1
Copper	Cu	63·5	Tellurium	Te	127·6
Fluorine	F	19·0	Thallium	Tl	204·4
Gold	Au	197·2	Thorium	Th	232·1
Hydrogen	H	1·0	Tin	Sn	118·7
Iodine	I	126·9	Titanium	Ti	47·9
Iron	Fe	55·9	Tungsten	W	183·9
Lead	Pb	207·2	Uranium	U	238·1
Lithium	Li	6·9	Vanadium	V	51·0
Magnesium	Mg	24·3	Zinc	Zn	65·4
Manganese	Mn	54·9	Zirconium	Zr	91·2

Four figure logarithmic tables

LOGARITHMS

	0	1	2	3	4	5	6	7	8	9	1	2	3	4	5	6	7	8	9
10	0000	0043	0086	0128	0170	0212	0253	0294	0334	0374	4	8	12	17	21	25	29	33	37
11	0414	0453	0492	0531	0569	0607	0645	0682	0719	0755	4	8	11	15	19	23	26	30	34
12	0792	0828	0864	0899	0934	0969	1004	1038	1072	1106	3	7	10	14	17	21	24	28	31
13	1139	1173	1206	1239	1271	1303	1335	1367	1399	1430	3	6	10	13	16	19	23	26	29
14	1461	1492	1523	1553	1584	1614	1644	1673	1703	1732	3	6	9	12	15	18	21	24	27
15	1761	1790	1818	1847	1875	1903	1931	1959	1987	2014	3	6	8	11	14	17	20	22	25
16	2041	2068	2095	2122	2148	2175	2201	2227	2253	2279	3	5	8	11	13	16	18	21	24
17	2304	2330	2355	2380	2405	2430	2455	2480	2504	2529	2	5	7	10	12	15	17	20	22
18	2553	2577	2601	2625	2648	2672	2695	2718	2742	2765	2	5	7	9	12	14	16	19	21
19	2788	2810	2833	2856	2878	2900	2923	2945	2967	2989	2	4	7	9	11	13	16	18	20
20	3010	3032	3054	3075	3096	3118	3139	3160	3181	3201	2	4	6	8	11	13	15	17	19
21	3222	3243	3263	3284	3304	3324	3345	3365	3385	3404	2	4	6	8	10	12	14	16	18
22	3424	3444	3464	3483	3502	3522	3541	3560	3579	3598	2	4	6	8	10	12	14	15	17
23	3617	3636	3655	3674	3692	3711	3729	3747	3766	3784	2	4	6	7	9	11	13	15	17
24	3802	3820	3838	3856	3874	3892	3909	3927	3945	3962	2	4	5	7	9	11	12	14	16
25	3979	3997	4014	4031	4048	4065	4082	4099	4116	4133	2	3	5	7	9	10	12	14	15
26	4150	4166	4183	4200	4216	4232	4249	4265	4281	4298	2	3	5	7	8	10	11	13	15
27	4314	4330	4346	4362	4378	4393	4409	4425	4440	4456	2	3	5	6	8	9	11	13	14
28	4472	4487	4502	4518	4533	4548	4564	4579	4594	4609	2	3	5	6	8	9	11	12	14
29	4624	4639	4654	4669	4683	4698	4713	4728	4742	4757	1	3	4	6	7	9	10	12	13
30	4771	4786	4800	4814	4829	4843	4857	4871	4886	4900	1	3	4	6	7	9	10	11	13
31	4914	4928	4942	4955	4969	4983	4997	5011	5024	5038	1	3	4	6	7	8	10	11	12
32	5051	5065	5079	5092	5105	5119	5132	5145	5159	5172	1	3	4	5	7	8	9	11	12
33	5185	5198	5211	5224	5237	5250	5263	5276	5289	5302	1	3	4	5	6	8	9	10	12
34	5315	5328	5340	5353	5366	5378	5391	5403	5416	5428	1	3	4	5	6	8	9	10	11
35	5441	5453	5465	5478	5490	5502	5514	5527	5539	5551	1	2	4	5	6	7	9	10	11
36	5563	5575	5587	5599	5611	5623	5635	5647	5658	5670	1	2	4	5	6	7	8	10	11
37	5682	5694	5705	5717	5729	5740	5752	5763	5775	5786	1	2	3	5	6	7	8	9	10
38	5798	5809	5821	5832	5843	5855	5866	5877	5888	5899	1	2	3	5	6	7	8	9	10
39	5911	5922	5933	5944	5955	5966	5977	5988	5999	6010	1	2	3	4	5	7	8	9	10
40	6021	6031	6042	6053	6064	6075	6085	6096	6107	6117	1	2	3	4	5	6	8	9	10
41	6128	6138	6149	6160	6170	6180	6191	6201	6212	6222	1	2	3	4	5	6	7	8	9
42	6232	6243	6253	6263	6274	6284	6294	6304	6314	6325	1	2	3	4	5	6	7	8	9
43	6335	6345	6355	6365	6375	6385	6395	6405	6415	6425	1	2	3	4	5	6	7	8	9
44	6435	6444	6454	6464	6474	6484	6493	6503	6513	6522	1	2	3	4	5	6	7	8	9
45	6532	6542	6551	6561	6571	6580	6590	6599	6609	6618	1	2	3	4	5	6	7	8	9
46	6628	6637	6646	6656	6665	6675	6684	6693	6702	6712	1	2	3	4	5	6	7	7	8
47	6721	6730	6739	6749	6758	6767	6776	6785	6794	6803	1	2	3	4	5	5	6	7	8
48	6812	6821	6830	6839	6848	6857	6866	6875	6884	6893	1	2	3	4	4	5	6	7	8
49	6902	6911	6920	6928	6937	6946	6955	6964	6972	6981	1	2	3	4	4	5	6	7	8
50	6990	6998	7007	7016	7024	7033	7042	7050	7059	7067	1	2	3	3	4	5	6	7	8
51	7076	7084	7093	7101	7110	7118	7126	7135	7143	7152	1	2	3	3	4	5	6	7	8
52	7160	7168	7177	7185	7193	7202	7210	7218	7226	7235	1	2	2	3	4	5	6	7	7
53	7243	7251	7259	7267	7275	7284	7292	7300	7308	7316	1	2	2	3	4	5	6	6	7
54	7324	7332	7340	7348	7356	7364	7372	7380	7388	7396	1	2	2	3	4	5	6	6	7

	0	1	2	3	4	5	6	7	8	9	1	2	3	4	5	6	7	8	9
55	7401	7412	7419	7427	7435	7443	7451	7459	7466	7474	1	2	2	3	4	5	5	6	7
56	7482	7490	7497	7505	7513	7520	7528	7536	7543	7551	1	2	2	3	4	5	5	6	7
57	7559	7566	7574	7582	7589	7597	7604	7612	7619	7627	1	2	2	3	4	5	5	6	7
58	7634	7642	7649	7657	7664	7672	7679	7686	7694	7701	1	1	2	3	4	4	5	6	7
59	7709	7716	7723	7731	7738	7745	7752	7760	7767	7774	1	1	2	3	4	4	5	6	7
60	7782	7789	7796	7803	7810	7818	7825	7832	7839	7846	1	1	2	3	4	4	5	6	6
61	7853	7860	7868	7875	7882	7889	7896	7903	7910	7917	1	1	2	3	4	4	5	6	6
62	7924	7931	7938	7945	7952	7959	7966	7973	7980	7987	1	1	2	3	3	4	5	6	6
63	7993	8000	8007	8014	8021	8028	8035	8041	8048	8055	1	1	2	3	3	4	5	5	6
64	8062	8069	8075	8082	8089	8096	8102	8109	8116	8122	1	1	2	3	3	4	5	5	6
65	8129	8136	8142	8149	8156	8162	8169	8176	8182	8189	1	1	2	3	3	4	5	5	6
66	8195	8202	8209	8215	8222	8228	8235	8241	8248	8254	1	1	2	3	3	4	5	5	6
67	8261	8267	8274	8280	8287	8293	8299	8306	8312	8319	1	1	2	3	3	4	5	5	6
68	8325	8331	8338	8344	8351	8357	8363	8370	8376	8382	1	1	2	3	3	4	4	5	6
69	8388	8395	8401	8407	8414	8420	8426	8432	8439	8445	1	1	2	2	3	4	4	5	6
70	8451	8457	8463	8470	8476	8482	8488	8494	8500	8506	1	1	2	2	3	4	4	5	6
71	8513	8519	8525	8531	8537	8543	8549	8555	8561	8567	1	1	2	2	3	4	4	5	5
72	8573	8579	8585	8591	8597	8603	8609	8615	8621	8627	1	1	2	2	3	4	4	5	5
73	8633	8639	8645	8651	8657	8663	8669	8675	8681	8686	1	1	2	2	3	4	4	5	5
74	8692	8698	8704	8710	8716	8722	8727	8733	8739	8745	1	1	2	2	3	4	4	5	5
75	8751	8756	8762	8768	8774	8779	8785	8791	8797	8802	1	1	2	2	3	3	4	5	5
76	8808	8814	8820	8825	8831	8837	8842	8848	8854	8859	1	1	2	2	3	3	4	5	5
77	8865	8871	8876	8882	8887	8893	8899	8904	8910	8915	1	1	2	2	3	3	4	4	5
78	8921	8927	8932	8938	8943	8949	8954	8960	8965	8971	1	1	2	2	3	3	4	4	5
79	8976	8982	8987	8993	8998	9004	9009	9015	9020	9025	1	1	2	2	3	3	4	4	5
80	9031	9036	9042	9047	9053	9058	9063	9069	9074	9079	1	1	2	2	3	3	4	4	5
81	9085	9090	9096	9101	9106	9112	9117	9122	9128	9133	1	1	2	2	3	3	4	4	5
82	9138	9143	9149	9154	9159	9165	9170	9175	9180	9186	1	1	2	2	3	3	4	4	5
83	9191	9196	9201	9206	9212	9217	9222	9227	9232	9238	1	1	2	2	3	3	4	4	5
84	9243	9248	9253	9258	9263	9269	9274	9279	9284	9289	1	1	2	2	3	3	4	4	5
85	9294	9299	9304	9309	9315	9320	9325	9330	9335	9340	1	1	2	2	3	3	4	4	5
86	9345	9350	9355	9360	9365	9370	9375	9380	9385	9390	1	1	2	2	3	3	4	4	5
87	9395	9400	9405	9410	9415	9420	9425	9430	9435	9440	0	1	1	2	2	3	3	4	4
88	9445	9450	9455	9460	9465	9469	9474	9479	9484	9489	0	1	1	2	2	3	3	4	4
89	9494	9499	9504	9509	9513	9518	9523	9528	9533	9538	0	1	1	2	2	3	3	4	4
90	9542	9547	9552	9557	9562	9566	9571	9576	9581	9586	0	1	1	2	2	3	3	4	4
91	9590	9595	9600	9605	9609	9614	9619	9624	9628	9633	0	1	1	2	2	3	3	4	4
92	9638	9643	9647	9652	9657	9661	9666	9671	9675	9680	0	1	1	2	2	3	3	4	4
93	9685	9689	9694	9699	9703	9708	9713	9717	9722	9727	0	1	1	2	2	3	3	4	4
94	9731	9736	9741	9745	9750	9754	9759	9763	9768	9773	0	1	1	2	2	3	3	4	4
95	9777	9782	9786	9791	9795	9800	9805	9809	9814	9818	0	1	1	2	2	3	3	4	4
96	9823	9827	9832	9836	9841	9845	9850	9854	9859	9863	0	1	1	2	2	3	3	4	4
97	9868	9872	9877	9881	9886	9890	9894	9899	9903	9908	0	1	1	2	2	3	3	4	4
98	9912	9917	9921	9926	9930	9934	9939	9943	9948	9952	0	1	1	2	2	3	3	4	4
99	9956	9961	9965	9969	9974	9978	9983	9987	9991	9996	0	1	1	2	2	3	3	3	4

	0	1	2	3	4	5	6	7	8	9	1	2	3	4	5	6	7	8	9
·00	1000	1002	1005	1007	1009	1012	1014	1016	1019	1021	0	0	1	1	1	1	2	2	2
·01	1023	1026	1028	1030	1033	1035	1038	1040	1042	1045	0	0	1	1	1	1	2	2	2
·02	1047	1050	1052	1054	1057	1059	1062	1064	1067	1069	0	0	1	1	1	1	2	2	2
·03	1072	1074	1076	1079	1081	1084	1086	1089	1091	1094	0	0	1	1	1	1	2	2	2
·04	1095	1099	1102	1104	1107	1109	1112	1114	1117	1119	0	1	1	1	1	2	2	2	2
·05	1122	1125	1127	1130	1132	1135	1138	1140	1143	1146	0	1	1	1	1	2	2	2	2
·06	1148	1151	1153	1156	1159	1161	1164	1167	1169	1172	0	1	1	1	1	2	2	2	2
·07	1175	1178	1180	1183	1186	1189	1191	1194	1197	1199	0	1	1	1	1	2	2	2	2
·08	1202	1205	1208	1211	1213	1216	1219	1222	1225	1227	0	1	1	1	1	2	2	2	3
·09	1230	1233	1236	1239	1242	1245	1247	1250	1253	1256	0	1	1	1	1	2	2	2	3
·10	1259	1262	1265	1268	1271	1274	1276	1279	1282	1285	0	1	1	1	1	2	2	2	3
·11	1288	1291	1294	1297	1300	1303	1306	1309	1312	1315	0	1	1	1	2	2	2	2	3
·12	1318	1321	1324	1327	1330	1334	1337	1340	1343	1346	0	1	1	1	2	2	2	2	3
·13	1349	1352	1355	1358	1361	1365	1368	1371	1374	1377	0	1	1	1	2	2	2	3	3
·14	1380	1384	1387	1390	1393	1396	1400	1403	1406	1409	0	1	1	1	2	2	2	3	3
·15	1413	1416	1419	1422	1426	1429	1432	1435	1439	1442	0	1	1	1	2	2	2	3	3
·16	1445	1449	1452	1455	1459	1462	1466	1469	1472	1476	0	1	1	1	2	2	2	3	3
·17	1479	1483	1486	1489	1493	1496	1500	1503	1507	1510	0	1	1	1	2	2	2	3	3
·18	1514	1517	1521	1524	1528	1531	1535	1538	1542	1545	0	1	1	1	2	2	2	3	3
·19	1549	1552	1556	1560	1563	1567	1570	1574	1578	1581	0	1	1	1	2	2	3	3	3
·20	1585	1589	1592	1596	1600	1603	1607	1611	1614	1618	0	1	1	1	2	2	3	3	3
·21	1622	1626	1629	1633	1637	1641	1644	1648	1652	1656	0	1	1	2	2	2	3	3	3
·22	1660	1663	1667	1671	1675	1679	1683	1687	1690	1694	0	1	1	2	2	2	3	3	3
·23	1698	1702	1706	1710	1714	1718	1722	1726	1730	1734	0	1	1	2	2	2	3	3	4
·24	1738	1742	1746	1750	1754	1758	1762	1766	1770	1774	0	1	1	2	2	2	3	3	4
·25	1778	1782	1786	1791	1795	1799	1803	1807	1811	1816	0	1	1	2	2	2	3	3	4
·26	1820	1824	1828	1832	1837	1841	1845	1848	1854	1858	0	1	1	2	2	3	3	3	4
·27	1862	1866	1871	1875	1879	1884	1888	1892	1897	1901	0	1	1	2	2	3	3	3	4
·28	1905	1910	1914	1919	1923	1928	1932	1936	1941	1945	0	1	1	2	2	3	3	4	4
29	1950	1954	1959	1963	1968	1972	1977	1982	1986	1991	0	1	1	2	2	3	3	4	4
·30	1995	2000	2004	2009	2014	2018	2023	2028	2032	2037	0	1	1	2	2	3	3	4	4
·31	2042	2046	2051	2056	2061	2065	2070	2075	2080	2084	0	1	1	2	2	3	3	4	4
·32	2089	2094	2099	2104	2109	2113	2118	2123	2128	2133	0	1	1	2	2	3	3	4	4
·33	2138	2143	2148	2153	2158	2163	2168	2173	2178	2183	0	1	1	2	2	3	3	4	4
·34	2188	2193	2198	2203	2208	2213	2218	2223	2228	2234	1	1	2	2	3	3	4	4	5
·35	2239	2244	2249	2254	2259	2265	2270	2275	2280	2286	1	1	2	2	3	3	4	4	5
·36	2291	2296	2301	2307	2312	2317	2323	2328	2333	2339	1	1	2	2	3	3	4	4	5
·37	2344	2350	2355	2360	2366	2371	2377	2382	2388	2393	1	1	2	2	3	3	4	4	5
·38	2399	2404	2410	2415	2421	2427	2432	2438	2443	2449	1	1	2	2	3	3	4	4	5
·39	2455	2460	2466	2472	2477	2483	2489	2495	2500	2506	1	1	2	2	3	3	4	5	5
·40	2512	2518	2523	2529	2535	2541	2547	2553	2559	2564	1	1	2	2	3	4	4	5	5
·41	2570	2576	2582	2588	2594	2600	2606	2612	2618	2624	1	1	2	2	3	4	4	5	5
·42	2630	2636	2642	2649	2655	2661	2667	2673	2679	2685	1	1	2	2	3	4	4	5	6
·43	2692	2698	2704	2710	2716	2723	2729	2735	2742	2748	1	1	2	3	3	4	4	5	6
·44	2754	2761	2767	2773	2780	2786	2793	2799	2805	2812	1	1	2	3	3	4	4	5	6
·45	2818	2825	2831	2838	2844	2851	2858	2864	2871	2877	1	1	2	3	3	4	5	5	6
·46	2884	2891	2897	2904	2911	2917	2924	2931	2938	2944	1	1	2	3	3	4	5	5	6
·47	2951	2958	2965	2972	2979	2985	2992	2999	3006	3013	1	1	2	3	3	4	5	5	6
·48	3020	3027	3034	3041	3048	3055	3062	3069	3076	3083	1	1	2	3	4	4	5	6	6
·49	3090	3097	3105	3112	3119	3126	3133	3141	3148	3155	1	1	2	3	4	4	5	6	6

ANTILOGARITHMS

	0	1	2	3	4	5	6	7	8	9	1	2	3	4	5	6	7	8	9
·50	3162	3170	3177	3184	3192	3199	3206	3214	3221	3228	1	1	2	3	4	4	5	6	7
·51	3236	3243	3251	3258	3266	3273	3281	3289	3296	3304	1	2	2	3	4	5	5	6	7
·52	3311	3319	3327	3334	3342	3350	3357	3365	3373	3381	1	2	2	3	4	5	5	6	7
·53	3388	3396	3404	3412	3420	3428	3436	3443	3451	3459	1	2	2	3	4	5	6	6	7
·54	3467	3475	3483	3491	3499	3508	3516	3524	3532	3540	1	2	2	3	4	5	6	6	7
·55	3548	3556	3565	3573	3581	3589	3597	3606	3614	3622	1	2	2	3	4	5	6	7	7
·56	3631	3639	3648	3656	3664	3673	3681	3690	3698	3707	1	2	3	3	4	5	6	7	8
·57	3715	3724	3733	3741	3750	3758	3767	3776	3784	3793	1	2	3	3	4	5	6	7	8
·58	3802	3811	3819	3828	3837	3846	3855	3864	3873	3882	1	2	3	4	4	5	6	7	8
·59	3890	3899	3908	3917	3926	3936	3945	3954	3963	3972	1	2	3	4	5	5	6	7	8
·60	3981	3990	3999	4009	4018	4027	4036	4046	4055	4064	1	2	3	4	5	6	6	7	8
·61	4074	4083	4093	4102	4111	4121	4130	4140	4150	4159	1	2	3	4	5	6	7	8	9
·62	4169	4178	4188	4198	4207	4217	4227	4236	4246	4256	1	2	3	4	5	6	7	8	9
·63	4266	4276	4285	4295	4305	4315	4325	4335	4345	4355	1	2	3	4	5	6	7	8	9
·64	4365	4375	4385	4395	4406	4416	4426	4436	4446	4457	1	2	3	4	5	6	7	8	9
·65	4467	4477	4487	4498	4508	4519	4529	4539	4550	4560	1	2	3	4	5	6	7	8	9
·66	4571	4581	4592	4603	4613	4624	4634	4645	4656	4667	1	2	3	4	5	6	7	9	10
·67	4677	4688	4699	4710	4721	4732	4742	4753	4764	4775	1	2	3	4	5	7	8	9	10
·68	4786	4797	4808	4819	4831	4842	4853	4864	4875	4887	1	2	3	4	6	7	8	9	10
·69	4898	4909	4920	4932	4943	4955	4966	4977	4989	5000	1	2	3	5	6	7	8	9	10
·70	5012	5023	5035	5047	5058	5070	5082	5093	5105	5117	1	2	4	5	6	7	8	9	11
·71	5129	5140	5152	5164	5176	5188	5200	5212	5224	5236	1	2	4	5	6	7	8	10	11
·72	5248	5260	5272	5284	5297	5309	5321	5333	5346	5358	1	2	4	5	6	7	9	10	11
·73	5370	5383	5395	5408	5420	5433	5445	5458	5470	5483	1	3	4	5	6	8	9	10	11
·74	5495	5508	5521	5534	5546	5559	5572	5585	5598	5610	1	3	4	5	6	8	9	10	12
·75	5623	5636	5649	5662	5675	5689	5702	5715	5728	5741	1	3	4	5	7	8	9	10	12
·76	5754	5768	5781	5794	5808	5821	5834	5848	5861	5875	1	3	4	5	7	8	9	11	12
·77	5888	5902	5916	5929	5943	5957	5970	5984	5998	6012	1	3	4	5	7	8	10	11	12
·78	6026	6039	6053	6067	6081	6095	6109	6124	6138	6152	1	3	4	6	7	8	10	11	13
·79	6166	6180	6194	6209	6223	6237	6252	6266	6281	6295	1	3	4	6	7	9	10	11	13
·80	6310	6324	6339	6353	6368	6383	6397	6412	6427	6442	1	3	4	6	7	9	10	12	13
·81	6457	6471	6486	6531	6516	6531	6546	6561	6577	6592	2	3	5	6	8	9	11	12	14
·82	6607	6622	6637	6653	6668	6683	6699	6714	6730	6745	2	3	5	6	8	9	11	12	14
·83	6761	6776	6792	6808	6823	6839	6855	6871	6887	6902	2	3	5	6	8	9	11	13	14
·84	6918	6934	6950	6966	6982	6998	7015	7031	7047	7063	2	3	5	6	8	10	11	13	15
·85	7079	7096	7112	7129	7145	7161	7178	7194	7211	7228	2	3	5	7	8	10	12	13	15
·86	7244	7261	7278	7295	7311	7328	7345	7362	7379	7396	2	3	5	7	8	10	12	13	15
·87	7413	7430	7447	7464	7482	7499	7516	7534	7551	7568	2	4	5	7	9	10	12	14	16
·88	7586	7603	7621	7638	7656	7674	7691	7709	7727	7745	2	4	5	7	9	11	12	14	16
·89	7762	7780	7798	7816	7834	7852	7870	7889	7907	7925	2	4	5	7	9	11	13	14	16
·90	7943	7962	7980	7998	8017	8035	8054	8072	8091	8110	2	4	6	7	9	11	13	15	17
·91	8128	8147	8166	8185	8204	8222	8241	8260	8279	8299	2	4	6	8	9	11	13	15	17
·92	8318	8337	8356	8375	8395	8414	8433	8453	8472	8492	2	4	6	8	10	12	14	15	17
·93	8511	8531	8551	8570	8590	8610	8630	8650	8670	8690	2	4	6	8	10	12	14	16	18
·94	8710	8730	8750	8770	8790	8810	8831	8851	8872	8892	2	4	6	8	10	12	14	16	18
·95	8913	8933	8954	8974	8995	9016	9034	9057	9078	9099	2	4	6	8	10	12	15	17	19
·96	9120	9141	9162	9183	9204	9226	9247	9268	9290	9311	2	4	6	8	11	13	15	17	19
·97	9333	9354	9376	9397	9419	9441	9462	9484	9506	9528	2	4	7	9	11	13	15	17	20
·98	9550	9572	9594	9616	9638	9661	9683	9705	9727	9750	2	4	7	9	11	13	16	18	20
·99	9772	9795	9817	9840	9863	9886	9908	9931	9954	9977	2	5	7	9	11	14	16	18	20

Index